STRANGE LABYRINTH

WILL ASHON was born in Leicester in 1969. Having worked as a music journalist, he founded the record label Big Dada Recordings in 1996, which he ran for over fifteen years, signing acts like Roots Manuva, Wiley, Diplo, Kate Tempest and Young Fathers. He also published two novels with Faber & Faber, *Clear Water* and *The Heritage*. He currently lives in Walthamstow, north-east London.

'An anarchic hymn to the scruffy edgeland of Epping Forest, the ancient wood that sits on the boundary between London and Essex' *Observer*

'*Strange Labyrinth* is a wonderful exploration of the tangled undergrowth of the psyche. Ashon is an anarchic Green Man; a puckish punk of the forests. He has invented a new genre: Gonzo Romanticism' Jon Day

'This book made me so happy. A mind-expanding journey, fiercely researched, expertly constructed and ultimately life-affirming' Luke Kennard

'From John Clare to Crass, Will Ashon unearths magic in a forest that is more than mere harbour and hide-out for dissidents, dreamers, poets and outlaws, but which represents an entire narrative strand of an ever-changing England. Here is deep questing into both person and place, masterfully delivered' Benjamin Myers

'Extensively researched and hilariously self-deprecating, *Strange Labyrinth* takes us on a journey that is funny, moving and fascinating' *Idler*

'I was hooked, enchanted even… [An] intensely personal, magpie's jewel box of a book… Extraordinary and entertaining... a splendidly erudite but engaging book. Ashon is a self-deprecating and discursive guide, often very funny… A wonderfully idiosyncratic, somehow very British book which delighted me from start to finish' *A Life in Books* blog

'Magnetic reading… There's a joy to the book's outrage against systemic control' *Minor Literatures*

'I found it mercilessly lucid, wildly expansive yet down-to-earth, and misanthropic as only books with real heart can be. There's a tendency to treat psychogeography as a form of archaeology but he bypasses anything resembling fossils for a more intriguing, irreverent and animated approach. These are fragments of the past brought to life in the present, and a fascinating, cynical yet wide-eyed and inspiring, despite itself, present set in the greater scheme of things. A journey into the dark and terrible maze that is England with a guide as much Minotaur as Theseus' Darran Anderson

'Combining expert research with wonderful self-deprecation, *Strange Labyrinth* is a brave and very funny account of modern anxieties' Paul Ewen

'Wilder than Macfarlane, funnier than Deakin and more emotionally engaged than Sebald, Will Ashon turns getting lost in the forest into high art, and great entertainment. By the end you'll probably be looking for a berth up a tree alongside him' Matt Thorne

STRANGE LABYRINTH

Outlaws, poets, mystics,
murderers and a coward in
London's great forest

Will Ashon

GRANTA

Granta Publications, 12 Addison Avenue, London W11 4QR

First published in Great Britain by Granta Books, 2017
This paperback edition published in Great Britain
by Granta Books, 2018

A CIP catalogue record for this book is available
from the British Library.

1 3 5 7 9 10 8 6 4 2

ISBN 978 1 78378 345 8
eISBN 978 1 78378 344 1

Typeset in Albertina MT by Lindsay Nash
Printed and bound by CPI Group (UK) Ltd, Croydon, CR0 4YY

In memory of Claire Tansley

In this strange labyrinth how shall I turn?
MARY WROTH

I'm not *against* the police; I'm just afraid of them
ALFRED HITCHCOCK

Contents

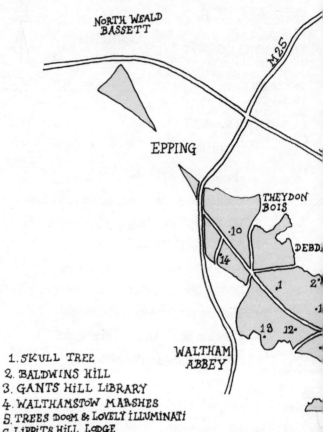

.7

NORTH WEALD
BASSETT

EPPING

M·25

THEYDON
BOIS

DEBD

·10

·14

·1

2·

·1

·13 12·

WALTHAM
ABBEY

SEWARDSTON

1. SKULL TREE
2. BALDWINS HILL
3. GANTS HILL LIBRARY
4. WALTHAMSTOW MARSHES
5. TREES DOOM & LOVELY ILLUMINATI
6. LIPPITS HILL LODGE
7. DIAL HOUSE
8. THE GREEN MAN
9. CLAREMONT ROAD
10. AMBRESBURY BANKS
11. HOLLOW PONDS

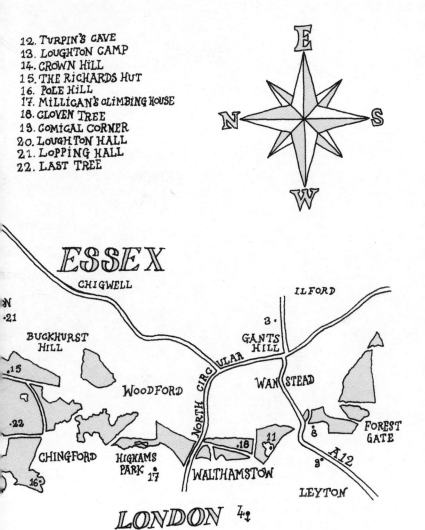

List of Illustrations

STRANGE LABYRINTH

CHAPTER 1

The Skull Tree | Baldwins Hill

I was looking for topiary when I found the Skull Tree.

It's perhaps some measure of how screwed up my priorities were at the time that I was wandering around a wood with that as my only pretext. I was having a crisis, at least I thought I might be, and I was using this task to hold that crisis at bay. It was all about providing structure, I suppose, the rules that I might operate within. Ordinary daily life is messy enough, but I'd chucked away all the routines, those points of famil-iarity which, like breadcrumbs placed on the forest floor behind me, I'd used to navigate through my existence. I'd set aside fifteen years of work and the status and self-definition I obtained from it. Having given it up, almost on a whim, topiary-hunting was what I was left with. It wasn't easy, but it was easier than figuring out what to do instead.

The Skull Tree was a beech, which hardly made it distinc-tive. There must have been tens of thousands of those through this particular stretch of land. It had been pollarded at some point, but not in the last hundred years or so. It was tall but no taller than the trees round it, maybe sixty feet or more.

It wasn't far from a busy road, but then nor was anywhere else in the forest. On the other hand, it wasn't close to any major path and I can only remember seeing anyone else near it on one occasion. Like me the passer-by was a middle-aged man without a dog. He looked slightly suspicious, sinister, my mirrored twin, so I suppose I did too. The beech was on a slope down towards a stream, although the streams here were more like ditches, what water they contained standing instead of flowing. A tree near to the ditch had blown down sometime in the last six months and remained enthusiastically alive, though recumbent.

My topiarian quest had been inspired by the sculptor Jacob Epstein. He'd been told that the monks who owned these woods until the Reformation had deliberately cut and bent the branches of the trees to make a menagerie of fabulous animals. Apparently he'd always been on the lookout for them. I could understand why the idea of living, growing monuments (of living, growing beasts) would appeal to a sculptor, but, perhaps owing to some lack of vision on my own part, I was struggling to spot them. I liked to think that was also true for Epstein. I'd decided he wasn't the most imaginative of artists, more of a technical whizz with a great pair of hands, the kind of creator who makes it look easy without ever finding a way to actually make it hard. And in all honesty, it's difficult to spot *anything* in a forest. There's always something in the way.

Born in 1880 on the Lower East Side of New York to Polish-Jewish immigrant parents, Jacob Epstein was your archetypal sickly child, his father complaining that 'all he did was draw and read'. I can only assume the totality of his considerable energy was being channelled from and to paper. By the time he was fourteen he was making enough from selling his pictures to rent his own studio, and by 1902 he'd saved up the money to buy himself a ticket to Paris, at that time the centre of the international art world. He studied there for a couple of years, then in 1904 he relocated to London, a much less cosmopolitan city where, he would later suggest, 'the bad air befogs' the residents' brains, resulting in their appalling aesthetic judgement. Although the sculptor claimed he was there to study the statuary at the British Museum, the main motivation seems to have been the presence of a Scotswoman called Margaret Dunlop, with whom he was living by 1906. The pair's relationship would last until Margaret's death in 1947, effectively stranding Epstein in our damp little kingdom of philistines. Within ten years of his wife's passing he'd been knighted, treading the well-marked path from enfant terrible to Establishment favourite.

These days Epstein is best known for a handful of works. *Rock Drill* from 1913–14 shows a Futurist-influenced robot humanoid attached to the gigantic spidery tool of its title, and has come to be read as a commentary on the industrialisation of war and, through that, of humanity itself. It was in fact a science-fictional prefiguring, and Epstein's decision to dismantle the sculpture can be seen as marking his break from the fascist and thanatophilic currents found in both English Vorticism and the Italian movement it made common cause with. *Jacob and the Angel*, from 1940–41, a staple of Tate Britain, is a massive carving in alabaster, both figures' legs like tree trunks, the fabled wrestle turned into a close embrace, with Jacob apparently relying on some form of passive resistance to occasion his escape. Despite the bulk of it, there's something in the angel's posture, in the massive, swooping tablets of its wings, which suggest that the pair could be a mile up in the sky in the blink of an eye. Day to day Epstein was more earthbound, making his living producing rather nice, rather facile bronze busts of the great and good, admirals and civil servants and politicians, realist pap lapped up by people who spent the rest of their time lambasting him as a degenerate.

Anyway, following Jacob's example I was looking for animal shapes in the trees, and the Skull Tree looked like an elephant. Or rather, it looked like the head of an elephant. Or more accurately still, the point at which the tree had previously been pollarded looked like a children's book representation of an elephant trumpeting his joy or anger to the sky, his ears actually forced upward, against gravity, by the power of his call. That is to say, it didn't look much like an elephant at all. Although I still somehow felt that it did, and this may have been as much to do with the smooth, pachydermal greyness of the bark as anything in its shaping. I was very keen to see elephants in the trees, had in some sense been predisposed to find them there.

Not being a great naturalist, or indeed any naturalist at all, once I got up close my main interest in the tree was the impact of people upon it. In particular, the carving on it. What's now known as Epping Forest was *saved for the nation* in 1878, in effect making it the UK's first national park. If you envisage Greater London as a rotated capital Q, where the inside of the 'O' is made by the North and South Circular roads and the outside by the orbital M25 motorway, then Epping Forest is the tail:

It is, in some senses, a simulacrum of a *real forest*, in that the activities you might traditionally have used it for are all banned: camping, having a fire, cutting wood, collecting mushrooms, catching rabbits—anything which might support subsistence living. Instead, under the terms of that government Act, it was envisaged as a leisure area for the working classes, a Cockney Paradise, somewhere for Londoners to spend quality time at the weekend getting fresh air and picnicking before returning to factories and smog and overcrowding and so on, the whole impedimenta of the Industrial Revolution. No surprise, then, that most trees within five hundred yards of an extant or former car park have been heavily cut into and carved. So systematically has this leisure activity been indulged in, it could almost be an industrial process itself.

To be frank, very few people seem to have devoted the full beam of their creativity to the task. The standard form is the algebraic equation of AB 4 XY or AB + XY, sometimes encompassed by a rudimentary heart. Elvis and the Sex Pistols get mentions on a tree at the Iron Age fort, Ambresbury Banks, right up at the top of the forest, not far from the town of Epping itself. Further south and across the busy road, which I could hear, I had found a couple of very old trees which had been marked up with a certain ghoulish fervour, a whole scribble of stuff. But my favourite, the Skull Tree, was wordless, almost unlettered. Instead, about four feet up from the ground, the trunk had a face carved into it, a bony head shaped a bit like an American muffin, the two eyes round knots in the wood, the nose a triangular dent, the mouth a weird, sliding moue. Directly above the head hung an upside-down T, a hammer dropping on the surprised skelly beneath. To the right, some initials, ballooned out by time and growth so that only an 'L' could be made out. It was masterfully done, perfect for anyone educated on pirate stories. The eyes were like marbles, a convex presence rather than the absence you'd expect from cutting, the contours of the bark blossoming out from the right pupil, the left one blazing up as if aflame or enhanced with false lashes, the final effect in combination with the mouth being one of either mild consternation or nuclear obliteration entering from stage left.

I wanted to know what the person who had carved the skull meant by it and, fearing the potential banality of a correct answer, I also wanted to make it up. I knew it couldn't be that old—I had read somewhere that after fifty years or so the bark of a beech tree would expand and change to such an extent that any gouges would disappear. Though, having said that, I had seen a carving dated 1887 for Queen Victoria's Golden Jubilee, and what kind of person would bother to scratch a false date on a tree? It never occurred to me at the time that the skull could represent some kind of a warning, the graphic equivalent of a Keep Out sign. I chose to view it as an invitation, not an omen. This was uncharacteristic.

There was a secondary reason, as well, for liking this particular beech. One hundred or so degrees clockwise round the trunk from the carving, the roots had formed a very flat, arse-sized platform at the tree's base, so that a wanderer could sit, weary, and eat his sandwiches without getting his jeans too wet. It was surprisingly comfortable in a straight-backed, ascetic kind of way. The ground around was covered in empty beechnut shells, the interiors soft and furry, so that if you stroked across them with your thumb they felt almost animal. Patches of lime-green moss poked through here and there. A squirrel at the top of a nearby tree stood rock-still watching me, coughing out a strange, bird-like warning to other creatures too well hidden by foliage for me to see.

I was to become very attached to my seat there, not least for the effort it took me to work out exactly where it was so that I could return and the subsequent pride in successfully being able to do so. Epping Forest is full of paths and tracks and most of them don't go anywhere. They meander round holly bushes, cut back on themselves, cross other paths endlessly, cross themselves, follow and don't follow the contours of the land as they see fit. They seem designed not to take you to a destination at all, but instead, in a kind of self-extinguishing folly, *to get you lost*. Especially in those early months, to find my way back to any place I had previously discovered was a pleasure in itself, satisfaction with a job well done which was sadly lacking from the rest of my existence.

The crisis. Not a hugely original one—part developmental, part existential, largely silly. I didn't even take it very seriously when I was in it. Its genealogy can be traced back at least as far as the start of the fourteenth century and Dante's *Divine Comedy*: 'In the middle of our life's path / I found myself in a dark forest'. In the footsteps of a million men and women before me, I was trundling along, deep into my forties, when, like Dante—though lacking the emollient of either Christian faith or personal genius—I found myself plunged into that dark forest. While Dante's was placed there by God, mine was of my own making, scrappier and more circumscribed as a result. I was both lost and trapped. I had no idea how I would spend the rest of my life, but I knew it had to be different to how it had gone before. My jaw ached the whole time. My eyesight seemed to be deteriorating, not in a way that could be measured by opticians, but so that close focusing, particularly where screens were involved, caused a fixed, migrainous floater to sit in the centre of my vision. My face, which I had once considered to be if not exactly *handsome* then at least open to the possibility of being handsome, collapsed into a thin-lipped, tired sourness. Late on a Sunday night my wife asked me, at the end of a long weekend of dispute and argument, *What do you want, Will? What is it you want?* Of course, I couldn't answer. I had no idea what I wanted.

Not exactly in response to the outlandish enquiry, but not exactly *not* in response, I quit my job. I convinced myself that my writing career—two novels published over half a decade earlier had met with mild critical indifference and almost no sales—had been harmed by the distraction of my day job, the jinx of certain people around me and the half-hearted depression which these factors encouraged in me. This day job was in a mouse hole of the music industry, a place I had made comfortable for myself over the years but which now imprisoned me in a cycle of faked optimism, hard-knock pessimism and ever-diminishing returns. I believed that the culture of the business had irrevocably changed. Also that the job I did was actually in some way indefensible, a kind of colonialism by stealth. Also—in some degree of contradiction, I'll admit—that it was a young man's game and, in that I was no longer young, unlikely to serve my financial interests in the medium to long term.

When you read all this you may think I'm indulging in some heady cocktail of self-pity and self-loathing and that I am inviting you to join in. I'm not. I want you to laugh at me. Often the 'I' staggering through the woods is 'lost' or confused or quite stupid. Please don't mistake him for me. I know exactly where we're going, although the route through the forest is as winding, complicated and contradictory as you would expect. He is my avatar out in the trees, a silent movie klutz, tall and pale with cartoon features—the pendulous ears, the pear-shaped nose, those heavy-lidded, bulging eyes shadowed beneath that extravagant brow—set in a cadaverous face. It's as if Nosferatu has gone to a fancy dress party disguised as Mr Potato Head. He is not as well dressed as I think I am. He is not as successful as I hoped I would be. He is neither as insightful nor as clever as I would like to be. I'm telling his story and he's acting out mine. We are locked together, actor-me and narrator-I, in order to deliver what follows, and we hate each other for it.

The crisis should have been over almost before it had begun. I had taken decisive action and my reward would be reaped with creative liberation. Never again would I waste my time on anyone's insecurities and neediness but my own. All the symptoms of my previous existence abated briefly and then, a few months later, reasserted themselves with gusto—the temper, the impatience, the face pain, the sour look, the drinking and so on. I had spent my life with too much to do and now I didn't have enough. The masterpiece I'd imagined writing was not forthcoming. I plugged gamely on because that's how I was, without any real sense of what I was plugging gamely on for. But some days even that mindless forward motion—words dribbled out onto my laptop with all the force and confidence of an old man peeing into a cup—deserted me and I left the house to fake it by wandering through the forest which passed in a thin strip of trees not that far from where I lived. I had read somewhere about a distinction between sitting writers and walking writers and it was so obvious which was supposed to be better that I decided to become one. Even though I wasn't writing anything, just walking. Pretending that I was a walking writer went some way to quelling my panic. I felt good about feeling bad as I trudged along, as if I were inhabiting the part more fully.

It was a short step from this pretence, barely a step at all, to telling my wife and children and then my friends and then distant acquaintances at parties and then even complete strangers that I was engaged in research for a book, a work of non-fiction, about the very trees I was hiding in, or at least about the other people who had hidden in them first. Having begun this fiction, though, I was forced to inhabit it, so that the difference between faking writing a book about the forest and actually writing a book about the forest became, almost instantly, very small. I began to spend my days in the British Library, calling up endless obscure titles, most long out of print, some of them driven down to London from stacks in a bunker up north, a whole forest of words printed on acres of felled trees, relating to this or that aspect of my *research*, which kept spreading and expanding in such a way that I could be sure it would never be complete. Gradually I convinced myself that, contrary to what I knew, this was a serious undertaking. I convinced myself, moreover, that this undertaking could, in and of itself, somehow save me.

If I ventured out into the forest to enact, inhabit or transcend my own crisis, I found plenty more waiting there to meet me. I came to see the place as a strange attractor for crisis, born out of crisis, both resistant to and complicit in a thousand more, creator and diffuser, as if its very character were revealed in this way. Or as if mine was, anyway. Take Jacob Epstein. We left him earlier, searching for trees that looked like animals. It was a crisis—a marital crisis—which brought the American artist turned British citizen out to the forest. And it was here that this crisis was played out, manipulated, magically warped into something else over the next twenty years. In pathology, a crisis is the turning point of an illness towards either recovery or death. So before we get there, we have to talk about the illness.

I'm not sure if there's a technical term for finding sexual partners for your husband but I do know that it must require a certain steely sense of priorities. This was the marriage of Jacob and Margaret. Apparently she didn't mind, confiding to her sister-in-law that her husband liked to do it in unusual and dirty ways, which obviously made her uncomfortable (possibly physically as well as morally). As a result, she left the rutting to the help, befriended the help, helped the help—all one big happy family. Her first daughter, Peggy Jean, born in Paris in 1918, was in fact the child of Meum Lindsell-Stewart, a model who lived with the Epsteins for a year or two before moving on, though the girl didn't find this out for a number of years. Peggy Jean was named after her adoptive mother, whom everyone called Peggy. To keep things from becoming too confused, I'll continue to refer to her as Margaret. As long as Margaret was in control of a situation, everything was fine.

That was how it worked until Kathleen Garman disturbed the equilibrium. In 1922, when Jacob was already forty-two and his wife forty-nine, he met Garman in a restaurant and started an affair with her behind his wife's back. The first Margaret knew of it was when she saw the post-coital bust Jacob made of his new girlfriend. Garman was twenty-one years old and recently arrived in London from the West Midlands, as keen

13

as her six sisters and sole brother to escape from their father, 'a cruel, bad-tempered, ferocious man'. Margaret had not been introduced. She invited Kathleen to come and have tea with her at the Epsteins' home on Guilford Street in Bloomsbury, and Kathleen accepted. Maybe she thought that the older woman was going to bless the relationship, or that she'd get a telling-off and an opportunity to defend the authenticity of her love. Instead, Margaret locked the door to the drawing room behind her young rival, took out a gun and shot her, hitting her in the left shoulder. Bang.

No one explains where Margaret got the pistol from, but they are happy to surmise that she meant to shoot the girl through the heart. It seems much more plausible that she hit her exactly where she wanted. She was a farmer's daughter, she knew how to get hold of a gun, how to load it, how to fire it, the two women couldn't have been more than a few steps apart. If she was aiming for the heart she would've hit it. It wasn't part of this performance to kill. Accounts vary as to what happened next. In one version, Garman, bleeding profusely, crawled out onto the street where her brother happened to be passing (or waiting?). In Margaret's own version, 'Kathleen lay wounded on the doorstep outside. Jacob said, "Margaret, take her in. You can't leave her out here. Let's take her in the house. She has no place to go." So I did.' Either way, Kathleen ended up in hospital. The doctors couldn't get all the lead out. It stayed there in her shoulder, causing her occasional pain and discomfort for the rest of her life.

You might think that when a man's wife has shot his lover, that's when he'll leave his wife. Or his lover. You're certainly expecting some kind of rupture. It's hard to tell whether Jacob made a decision at this point or failed to. Sometimes the two are very close. He turned up at the hospital and begged Kathleen not to press charges. We can only speculate how she felt as she lay in bed with her shoulder bandaged up while he told her that 'I must protect my wife'. On the other hand, it was after this that Epstein established a second household in Regent Square, less than ten minutes' walk from home, visiting Kathleen there every evening after he'd finished work and staying over on Wednesday and Saturday nights. Margaret, meanwhile, invited Kathleen to ride around Hyde Park with her in an open cab so that the world would know some kind of truce had been reached. This wasn't the end of the crisis, more like its beginning. Margaret Gilmour Dunlop, *Mrs Epstein*, was enacting some arcane rite that only she understood.

Margaret looms in the background of the biography of Epstein, a hazy presence or prominent absence, the long shadow on the floor. We know she was Scottish, everyone agrees about this. It's information that places you on the very lowest rung of the biographical ladder. One account has her growing up on the Isle of Mull among a full kirk of Presbyterians, another on a farm in Ayrshire, south-west of Glasgow. This source suggests she married a minister, that source a lord. The most plausible story is that Mr Thomas Williams was a barrister. Or a civil servant. And that Margaret worked for the Post Office. Or was a journalist. If it's hard to imagine that a Calvinist minister would have been in Paris lunching with the renowned Belgian anarchist Victor Dave, as Margaret and her husband were in 1903 when they met Epstein, the same is probably true of two employees of the Post Office. And how had she escaped from a farm in Scotland and made it to London in the first place? In his autobiography, Arthur

Ransome, best known as the author of *Swallows and Amazons* but a bohemian in his youth, recalls regularly visiting the couple at their fifth-floor flat on Gray's Inn Road. Williams was 'a compact little Welshman with a fine voice' and his wife a 'peasant lass who had educated herself'. He continued to call upon them until 'Peggy visited Paris and there met a sculptor as remarkable as herself'. We know she was a committed communist, described after her death by the composer Cecil Gray as 'a Red of the Reds'. But this communism was less the scientific Marxism of twentieth-century dogma than a kind of faith. In 1937, when Peggy Jean was to marry a young American doctor, Norman Hornstein, Margaret wrote to her sister-in-law, Sylvia, that 'at first I was afraid he was too beautiful, but Communism has been his religion, and that has kept him good I think'. And there is something touching in that. Maybe she was naive. Maybe I am, too.

While it might be hard to find out where a woman comes from or what she does for a living, there's never a problem if you want to know about her appearance. Sylvia, Jacob's younger sister, who lived in the US and first met Margaret in 1927, remarked that she 'looked just like a Buddha. She had hennaed hair and lovely brown eyes...everything was heavy on her. She had a heavy face with big jowls...She was heavy... a presence.' Cecil Gray, a fellow Scot and very much part of the bohemian set around Bloomsbury (he fathered a child with Hilda 'H.D.' Doolittle, who lived just round the corner from the Epsteins), said she was 'carved like a monumental statue', as if she were somehow Epstein's creation rather than he being hers. For a few years, Dolores was Epstein's live-in model, apparently one of Margaret's early efforts at tempting Jacob away from Kathleen. The mononymous sitter later sold a supercharged version of her life story to William Randolph Hearst's *American Weekly* magazine under the marvellous title 'The Fatal Woman of the London Studios' (they obviously didn't trust their readers to carry on if they used any French). In one instalment she described Margaret as 'a largely built Scotchwoman with a mop of untidy flaming red hair, turquoise stars in her ears, wearing men's golfing shoes, knitted woolen stockings and a sleeveless black dance frock'. Another biographer helpfully explains that by 1918, when Margaret was forty-six, she was already suffering from 'the dreadful ailment which lowers the metabolism and thickens the skin, and this changed her from the good-looking redhead to a person of gross appearance'.

Perhaps the most striking description of all comes from Maria Hollingsworth. Hollingsworth met the Epsteins when they visited her aunt in order to choose a puppy for their young son, Jackie, in 1936: 'Peggy sat like a menacing Buddha and said very little while dominating the whole room...her head was hennaed bright red and her legs so swollen with dropsy they looked as if they were going to burst. She breathed and walked with great difficulty but showed great tenderness to the child [Jackie]. I thought her terrifying, a primitive goddess, silent yet powerful. If the words "old soul" mean anything at all, Peggy was one, very old.'

I came across a photograph of Margaret and her future husband. Taken in Paris in 1903, around the time the couple met, it shows the two of them with Epstein's friend Bernard Gussow and his wife-to-be. Margaret sits in profile looking across towards *Jacques*, as she always called him. She doesn't resemble the Buddha, though her smile has a certain inscrutable quality. It's a long, strong face, with high, round cheeks and a gently curving nose. Her neck is long too, thin and graceful. There's a sense of complicity and excitement in the picture. Everyone is looking at Epstein as if he's just broken wind or made an off-colour joke. Only Margaret seems to be indulging him. She has one hand in her lap and with it she is making the sign of the horns, the index and little finger extended, the middle two folded over.

Nobody's hand rests like this naturally or by accident, so it's reasonable to assume Margaret is signalling something. It was only in the 1970s that the gesture came to have satanic overtones, so it's nothing of that sort. Is she hinting that she is about to cuckold her husband? Is she warding off the evil eye? Is this the Karana Mudra, as practised by the aforementioned Buddha? One thing is certain: Margaret was a confident manipulator of symbols, which is another way of saying she was an artist. Having no other canvas to paint on, no other rock to chisel at, she used her husband and his associates. Having no tools, she used performance and ritual, making those around her actors in a drama only she could understand. Having no stage, she used the world, she used everyday life. And, in time, she came to use the forest—the perfect terrain for anyone interested in discomposition, the swapping of roles, in enchantment and some form of malign, retributive magic.

In 1922, shortly after the rivals' trip around Hyde Park, Margaret woke model/lodger Dolores early one morning and told her they were going house-hunting: 'Put a cloak over your nightdress, tie a scarf around your head and slip on your golden sandals…We mustn't wait a moment.' The pair set off for Essex, and by the end of the day had rented a small cottage in Baldwins Hill as a weekend retreat and occasional workspace for Jacob. The Epsteins were to keep a property here for the next thirty-odd years. Many of Jacob's most famous sculptures were begun, worked on or completed in the shed at number 45 or, later, number 50, a slightly grander house called Deerhurst. He was also to paint hundreds of canvases of the woods, but I'll come to those in a while.

Baldwins Hill sits right up at the back of the town of Loughton, now best known for its appearances in TV show *The Only Way Is Essex*, in which orange young people have pretended to be themselves for over a decade. For now, I'm going to use Mansion House in the City of London as a trig point to orient ourselves by. Mansion House is the official residence of the Lord Mayor of London, and headquarters of the Corporation of the City of London, who manage Epping Forest. It lies roughly two miles north and east of the Houses of Parliament and, if that building is an ancient symbol of democracy, Mansion House is an equally potent symbol of money and commerce. Loughton, outside Greater London in the beginnings of Essex but still on the tube line, lies approximately thirteen miles north-north-east of Mansion House. I live in the south-eastern corner of Walthamstow, about halfway along the line between the two, down towards the southern end of the forest.

Baldwins Hill is some kind of nexus. Tucked away behind the more picturesque clapboard cottages of Little Cornwall, it's made up of a slightly scrubby mix of old and new housing developments whose inhabitants can lay claim to a long history of oppositionalism, independence and bohemian tendencies. No other settlement near Epping Forest has as many connections to the people we're going to meet in this book.

This is in part because of the access it grants to the forest itself. The widest, densest section of woods runs from Chingford (ten miles north-north-east of Mansion House) up to Epping (sixteen miles north-north-east) and Baldwins Hill is on its eastern lip about halfway along that stretch. Here, the forest is at its deepest and most bewildering, so that you feel both near to its centre and that the idea of a centre is ridiculous. If you shuffle off the muddy path behind Deerhurst—the second of the properties the Epsteins rented—you immediately find yourself in a landscape of disorientation. Wet clay slides out from under your feet. Holly, which tends here to a kind of straggling, threadlike disbursement, snags at your trousers, at your jacket, at your head. You walk up a steep slope and find yourself next to a pond, its surface dark and cold. You walk down another steep slope and emerge onto a mossy summit. The top stone of a small cairn of flints has been written on and it states that this is Court Hill.

I only found Court Hill because there's an inclination when walking under trees to head towards the patches of light. Even if you try not to you find that you are. The sun marks out routes which otherwise aren't there. The first time I came across this summit was also the first time I climbed a tree in the forest, scrabbling up a rotten stump of shaking wood with only one live branch in the hope of seeing out over the canopy. I was still in mountain mode back then, looking for vistas instead of enjoying the interiority afforded by blundering between trees. The almost-dead beech shook beneath me and I scrambled down as soon as I got up, disappointed by my failure to *see anything*, that sense that whatever direction I looked in was the wrong one. When I left the top of the hill along a path heading north, I was sure I was ascending.

The standard explanation for Margaret's renting the cottage in Essex is as confusing as the landscape. Epstein's biographers suggest she was trying to separate Jacob from Kathleen, but Baldwins Hill was only twenty minutes from Bloomsbury by train. It didn't exactly represent a gulf over which devoted lovers couldn't jump. Although Kathleen doesn't appear to have met him there, he arranged a spot in the woods where he would rendezvous with one of her sisters in order to exchange letters when he stayed there for more than a day or two. It was in this way that he heard about the birth of his first son, Theodore, in 1924. The bullet wasn't meant to kill Kathleen and the cottage in the forest wasn't meant to stop the relationship. Margaret called Kathleen the Witch of Mecklenburgh Square, and it seems that what she feared most in her rival was what she saw of herself, dressed in black, struggling back up the steps of her house, the keys to the door on a chain round her waist. What the pair needed was a suitable terrain on which to play out their conflict. Take a moment to examine those words from Dolores's highly coloured account: *Put a cloak over your nightdress, tie a scarf around your head and slip on your golden sandals.*

The writer Sara Maitland notes that in the 1857 edition of *Grimm's Fairy Tales*, over half of the 210 stories are located in the forest and another tenth of them have forest themes or images. Think of 'Hansel and Gretel' or 'Snow White', think of 'Beauty and the Beast' or 'Rumpelstiltskin'. The American academic Robert Pogue Harrison details the ways in which notions of the forest have fed into and affected Western culture across his magnificent book *Forests: The Shadow of Civilization*. In the Arthurian romances of Chrétien de Troyes, Lancelot goes mad in the forest not once but four times. In the ancient

Welsh text the *Mabinogion*, Perceval is raised in the forest in total isolation from humanity. In the Mesopotamian legend of Gilgamesh, the hero travels to Cedar Mountain to kill the Huwawa, the forest demon or guardian (depending on how you look at it). In Shakespeare, lovers wander lost in the forest and forget themselves and who they are, or the forest marches forward and Macbeth knows he is doomed. The Greek goddess Artemis hunted through the forest and when Actaeon saw her naked, she turned him into a stag and watched as his own dogs ripped him to pieces. At Nemi the priest of Diana was always an escaped slave who dared to pull down a golden bough from one of the trees in the sacred grove and then kill the previous priest, himself an escaped slave who had in turn pulled down a golden bough and killed his predecessor. That golden bough was first used by Aeneas—Virgil's mythical founder of the Roman Empire—in order to gain entry to the underworld, the land of the dead.

The forest has lost none of its power in our own time, though we have forced our fears onto minors. From *The Hobbit* to *Wind in the Willows*, it lurks at the edges of children's nightmares. At Hogwarts, of course, a prototypical Forbidden Forest pushes up on the edges of the school's grounds, the border beyond which magic becomes not good or evil but wild, savage and uncontrollable. What is it about trees that makes us fear them when they bunch up around us? Size and age must play a part, making toddlers of us all. Stillness, but also movement, the way a wood seethes with activity but never quite in front of you. The lack of clear sightlines, the endless places for people or animals to hide and jump out at you. The disorientation of what can seem like an infinite repetition rather than countless variations too tiny or recondite to be picked up by your eyes. Darkness, shadows, an occluded sun. Does it really need explaining? If you've ever spent any time out in the woods on your own you will have felt it, if only for a moment, even in that apparently tame, fake forest where I made my perambulations.

Margaret led her husband here, out among these trees, to be bewitched. I don't mean that the sculptor found the forest enchanting or pretty. I mean that he was taken over by it, caught in a compulsion beyond his control, pulled away from his own vision of himself, a cobbler trapped on a production line making cheap shoes designed only to break, or a fairy-tale prince whose legs won't stop dancing. Describing his days out in the forest with Peggy Jean in the summer of 1933 Jacob wrote, 'I would go out with my daughter and we did not have to walk far before seeing something worth painting. As usual with me, what started as a mere diversion became in the end a passion, and I could think of nothing else but painting. I arose to paint and painted till sundown.' He produced up to one hundred canvases of the trees, then found himself painting still-lifes of bunches of flowers, not one or two but hundreds of these as well, less an artistic decision than a compulsion, a curse. These flower paintings and forest paintings signalled the end of any sense that Epstein was a modernist, stripped the last claims of radicalism from him. Ironically, though, by producing something so eminently crowd-pleasing, by packing them from floor to ceiling onto gallery walls, by selling some of them to an American company to be printed onto tablecloths, he confirmed himself to the British press as the quintessential modern artist: an unprincipled fraud.

Not that the British press needed much encouragement to attack Epstein. One of the first pieces completed out in the shed at Baldwins Hill during the long, cold, foggy winter of 1924, when the wind hammered in against the thin wooden walls, was his monument to the naturalist T. H. Hudson, named *Rima* after the central character in Hudson's book, *Green Mansions*. The work he created can still be seen in Hyde Park, tucked away on the south side of a plant nursery across the road from the Serpentine Gallery. It's a quiet spot, not much visited, with a couple of benches sat in the shade of the trees, possibly the most controversial bird garden in the history of art. At the front is a rectangular pond, behind that a carefully mown lawn and, right at the back, a stone monument listing the years of Hudson's birth and death in Roman numerals. Between them is mounted Epstein's relief.

It shows Rima, the woman of the forest, naked, her hair like a flag over her right shoulder, her two arms held out and raised from the elbow, about to attempt flight. The space around her is filled with birds—big, rather ugly, abstracted, symbolic shapes, as if the best of her feathered friends are undersized dodos. There is nothing remotely pornographic about the work. Nor is it aesthetically or theoretically challenging, unless your theory is that all art should be of cute dogs playing snooker. If anything, it hints at the influence of Epstein's long study of the antiquities in the British Museum.

The monument, unveiled by Prime Minister Stanley Baldwin in May 1925, was met with outrage. The *Daily Mail* ran the headline: 'THE HYDE PARK ATROCITY – TAKE IT AWAY'. By November of that year it had been vandalised for the first time, on this occasion with green paint spelling out *Oh! Epstein!* (I particularly like that first exclamation mark). The reaction of the *Morning Post* was to state that 'the inevitable has happened' and hence to start a campaign not for the graffiti's removal but for the sculpture's. Over the next ten years it was vandalised a further three times. In 1929 it was tarred and feathered. In 1935 chemicals were poured on it which began to burn into the rock when it rained.

None of this was a reaction to the art. It was a reaction to the fact that Epstein was Jewish. Look at the work and it's quite hard to get a grip on what's going on. Read the headlines and it's a little easier. The *Daily Express* met Epstein's sculpture *Genesis* (1929–30) with the headline 'MONGOLIAN MORON THAT IS OBSCENE', and the openly fascist papers were far worse: 'The intensity of revulsion aroused by them cannot be explained merely on the grounds of their semitic cast.' The *Observer* described one of his exhibitions as 'unutterably loathsome'. Earlier, a certain Father Vaughn wrote in the *Weekly Graphic* that Epstein's *Christ* looked like 'some degraded Chaldean or African, which gave the appearance of an Asiatic-American or Hun-Jew'. Epstein was an outsider by dint of birth, not because of anything he made.

Interestingly, his Jewishness seems to have been part of his appeal to Margaret. A connoisseur of the symbolic, she drew a parallel between the Israelites and the Scots. In a letter to Sylvia regarding Peggy Jean's future, she claimed her adoptive daughter would be fine as soon as she met and married 'some nice doctor, Jewish-Scotch or Jewish or Scotch'. And indeed the Declaration of Arbroath, from 1320, which asserted to the Pope the independence of that nation from England, also claimed that the Scots people came originally from Greater Scythia and measured their longevity in terms of Moses and the Red Sea. It was only after Jacob and Margaret met that the Lower East Side boy who had rejected the religion of his parents began to return, again and again, to the stories of the Torah for subject matter. *Adam* and *Jacob and the Angel* (originally refused by the Tate when it was offered to them) ended up being displayed in a sideshow in Oxford Street with a sign saying that visitors would get their money back if they weren't suitably shocked and horrified. Around the same time (the early 1950s!), *Consummatum Est* and *Eve* were to be found in Blackpool as part of a display including 'human heads shrunk to the size of apples by Indian head-hunters, moving marionettes and the embalmed body of Siamese twins'.

B ack to the bewitchment. It may just have been a midlife crisis Epstein had been storing up, his personal dark forest, but I prefer to see something else there, something mulchier and less tedious. Eleven years had trundled by since Kathleen Garman and Jacob had begun their relationship. Margaret had been nothing if not patient. In the interim, Kathleen had borne the sculptor three offspring: Theodore in 1924, Kitty in 1926 and Esther in 1929. Margaret never acknowledged these children, nor made them welcome in her home. In the years immediately after her own death two of them would die, one of a heart attack in the back of an ambulance, the other committing suicide. In the summer of 1933, Jacob and Kathleen had some kind of an argument and she left for two months to visit one of her sisters, who was living with a fisherman in Martigues, north-west of Marseilles. Jacob started wandering out into the forest to paint. He painted and painted, from morning till dusk. He couldn't stop. He justified it after the fact as his usual *passion*, but it wasn't. And one aspect of his account is incomplete. He didn't just take Peggy Jean with him. He took Isabel Nicholas, his new model.

Nicholas vanishes from the story almost as quickly as she enters it, and maybe that's because of the way her own narrative echoes Margaret's—which is another way of saying it's because she's a woman. An accomplished, thoughtful painter (usually recognised under her third husband's name, Rawsthorne), whose abstracted figuration and use of light recall both Francis Bacon and Turner, she's better known as the former's model and muse, as the lover of Giacometti who sat for Picasso, rather than as an artist in her own right. Some of the Epstein biographies don't even get her name right. It's hard to talk about the bust the sculptor made of Isabel Nicholas without becoming complicit in this narrative. It certainly isn't one of his best pieces, although in that it displays his primary interest in her, it is at least honest. To put it bluntly, Epstein seems to have devoted more attention to sculpting Isabel's tits than her face. And yet...their balloony, cartoonish aspect, their almost supernatural uplift, the fact that they look more sentient than their owner, all suggest something else, some unease, an anxiety.

Jacob wrote to Kathleen in Martigues, desperate, pleading letters: 'I can only regret this parting & until I see you cannot be happy.' He occupied these sad, difficult times by having sex with Isabel. What's less clear is if he wanted to, or simply felt compelled to. They did it on Isabel's bed in the room she shared with his daughter, Peggy Jean, on one occasion leaving his sock for her to find. They must have done it out in the woods too, on the days they left Peggy Jean at home. And then the canvases, the hundreds of canvases, as excuse, as cover, as diversion, as justification. All he can see is trees: beech trees, hornbeam, oak, their limbs tangling behind his eyes. That was all I could see too, but I was on my own, walking his haunts, down from Great Monk Wood and its elephant trees to the oddly bleak Blackweir Pond, searching for the mound where he'd wanted to put his unfinished sculpture, *The Visitation*, and finding only a discarded fisherman's umbrella, the trees here scarred with writing, a thick incomprehensible scribble suggesting that even in 1933 it wasn't a secluded enough spot for al fresco sex.

I was guilty, of course, of a category error, based on the assumption that people don't like an audience when they're copulating. If Epping Forest is known for anything these days, it's dogging. When I told people I was writing a book about the area it was one of the first things they mentioned. Later a Verderer—a forest officer, elected to represent the interests of *commoners*—would outline for me the favoured dogging sites for heterosexual and homosexual couplings and the terrible problems they caused in terms of littering. There has even been a TV series devoted to the phenomenon. I watched an episode and was scared of and depressed by nearly everyone who appeared in it. Dispiritingly, the women all seemed damaged, manipulable, battling with a vacuum of self-esteem. The men were needy, sad and yes, manipulative, playing out porn fantasies in the woods behind their houses, indulging in a long, sorry story about their own irresistibility. I tried to take them on their own terms—that they were just less repressed than me—but I wasn't that open-minded. A man called Les claimed that he first heard the term 'dogging' back in the 1980s when some men surrounded his car and began masturbating while he and his girlfriend were shagging on the back seat at High Beach, a famous Epping Forest beauty spot on the western flank of the woods across from Baldwins Hill. It surely goes back much further. Maybe Jacob and Isabel drew a crowd. If they did, I'd bet that Margaret summoned them, hooded, participants in a rite.

I don't suppose Epstein ever bothered with condoms, which would at least have helped with the littering problem. By the start of 1934, Isabel was pregnant. As she had with Peggy Jean sixteen years earlier, Margaret decided she would be the mother of the new baby. This must have been, in some sense, what she'd been waiting for, her revenge on Kathleen Garman. With the help of Peggy Jean she attached pillows beneath her dresses and, sixty years old, pretended she was pregnant. Considering that no one would've believed this, we're left with the thought that she was engaged in another performance here, playing a part for reasons of her own, a kind of enact-ment of motherhood, a way to make the mundane mythic, her husband's rutting an act of the gods. In September Jackie Epstein was born. Isabel moved to Paris to further her career as an artist, possibly exiled by her child's surrogate. Margaret returned home with her new baby boy. Kathleen only found out that Epstein had fathered another child when a mutual acquaintance mentioned it a further nine months later. She was devastated. Margaret's feelings go unrecorded. But this is the moment when, through Jacob's latest, sanctioned infidelity, she emerged from the crisis of Jacob's perfidy and reasserted her central role in his life and art. Shortly after, Epstein began work on *Consummatum Est*, another gigantic alabaster sculpture showing Jesus lying on the ground following his crucifix-ion, his head raised as if to speak, his palms upturned, solid, heavy, as if his body imprisons him. The piece was named after Christ's supposed last words: it is finished.

Traditionally, in fairy tales and horror movies, there is a cost to casting spells, to getting your own way, particularly if you happen to be female. Child of an infertile woman who acted as mother to everyone, an inscrutable, primitive goddess, Jackie Epstein was the result of a cushions-up-the-jumper virgin birth. Born of the trees stretching out and away behind their house, you would expect him to be a stag-horned protector of nature, a green man with leaves growing from his mouth and across his face, literally a Jack o' the Green. Instead, 'curiously enough', he was 'only interested in machinery and ships…Engines are his…heart's delight. He thinks of a sub-marine as a paradise.' He would go on to be a racing driver and mechanic, a man obsessed with cars and combustion engines, as though Margaret's magic had released a power into the woods she couldn't understand.

In 1947, Margaret herself went out like a witch, dressed in her familiar black, her chain round her waist, slipping on the icy steps and cracking her head after she'd paid for the cab back from Essex. In their house at Chelsea was a locked room full of *treasures* she'd bought at auction. With her death it had turned to junk—costume jewellery and bits of tat, pissed on by her horde of cats. Kathleen—her enemy, her reflection, her vanquished rival, her successor—arranged for the whole lot to be thrown away. Jacob was also left diminished, tatty. The era of his large-scale stone pieces, those smooth, heavy, powerful emanations of his wife's magic, was over.

I sat beneath the Skull Tree for a long time on the day I discov-ered it, even after my sandwiches were gone and dampness had begun to soak its way into my personality. Eventually, starting to feel cold, I got up and explored the surrounding area, less a stag marking out its territory than a traumatised poacher checking for non-existent traps. I made my way over to the Skull Tree's fallen neighbour, up near what had been its crown, now pointing away along the ditch, feeling that odd, vertiginous sense of imbalance that toppled trees transmit to their surroundings. It had collided with another beech as it went down, the force breaking its highest branches so that they cascaded around the other trunk to make a rudimentary wigwam. Having checked for potential witnesses to my stupidity, I pulled back some foliage, ducked under the outer branches and waddled forward, my calves wedded to my thighs, until I found myself in a green mansion, or at least an antechamber, a private room to crouch in. For a moment, fondly, I imagined using this as an office, an HQ for my meandering, away from rain and prying eyes. Then the heat, the humidity and the flies all struck me, those small ones that move as if mapping coordinates in three dimensions before lodging themselves in the corner of your eye. I staggered out again, brushing spiders with filament legs from my hair, sneezing from the build-up of tree pollen and sap vapour. Standing there, I looked back at the Skull Tree. From this angle, perhaps from any angle, it looked absolutely nothing like an elephant at all.

CHAPTER 2

Walthamstow Marshes | Gants Hill Library

I was crouching next to a pond hoping that the spirit of
Ken Campbell would somehow *animate me*. I wasn't feeling
particularly animated but then it wasn't much of a pond. It
wasn't that deep or even that watery. It was choked with weeds
and bulrushes and, considering that it was in the middle of
Walthamstow Marshes, it looked more like an extension of its
surroundings—an intensification, perhaps—than something
deserving of its own separate category. The cows sometimes
kept in the field had made its borderlines more fluid still, the
hoofprints leading to the water's nominal edge now themselves a succession of miniature ponds, joined together into
a wider zone of *almost-pond*. In fact, as I looked round I realised that this almost-pond filled the whole field and indeed, as
that field was my whole world right then, the whole world. I
watched two crows involved in a tussle which, as with drunks
in a club, could've been dispute or courtship. A train moaned
past on the raised tracks behind me. The rain was thin and
diffuse, more a way of thinking than precipitation. My feet
were soaked so that it was hard, even, to tell for sure where

they ended and the pond began. I tried to imagine what it must have looked like here in 1945, when the V2 rocket had just made the crater, already filling with water but still unmistakably a disruption of the landscape rather than a feature of it. I couldn't manage it—there was too much to subtract as well as too much to add. The bench and table once installed here were gone, anyway, returning the area to the kind of soggy messiness—all dead thistles and smog-smudged skies—which Ken Campbell would first have found.

Even if you didn't know his name you'd probably recognise Ken. A short, pot-bellied bald man with a hooked nose and explosive eyebrows, which he kept untrimmed in order to be certain of picking up TV gigs as a mad scientist, he was Alf Garnett's neighbour, Fred Johnson, in the long-running sitcom *In Sickness and in Health*, made a cameo appearance in *Fawlty Towers* and fronted a number of popular science series for Channel 4 in the late 1990s. In a perhaps unique straddling of popular entertainment and the avant-garde, he also had roles in films by Derek Jarman and Peter Greenaway, although he acts just the same in all of them, as if he's both surprised and rather pleased by how *crap* it all is, and is hoping to make it a little worse. Mike Leigh, the film director who knew him from the pair's time at RADA, described Ken as 'the outsider's outsider', which is something like winning the Nobel Prize for outsiderdom. Though it would be even better if Mike Leigh actually was a gentleman of the road instead of only looking like one.

In the late 1980s, Ken would take his dog out for a walk on these marshes every morning. The mutt was named after Werner Erhard, whose confrontational *est* workshops were all the rage in the 1970s. Ken was such a fan of Erhard's he even paid for John Sessions to attend a course, an experience which precipitated something close to a nervous breakdown in the young actor. Ken and Werner would cross the river and make their way to this pond, where the dog would dive in and try to wrestle rubbish from the sludgy shallows. Ken would stand and wait for Werner to emerge. These days the marshes—sandwiched to east and west by accelerating gentrification—bustle with people at the weekend, the river spangled with hearty-looking folk training for the rowing club, sculling past motley houseboats moored, shabby, along the banks now that London land is too expensive to live on; joggers in a range of high-performance Lycra huffing by on the paths beyond. Back then the place was emptier, scruffier and wilder, so that Ken amused himself by pretending to be Tarkovsky's Stalker edging through the Zone. The scene repeated itself for months. Ken had nothing better to do. Earlier career highs had given way to a long hiatus, and he had no work.

Eventually the council put up a picnic bench and table next to the pond. Rather irritated by this at first (it put the kibosh on his *Stalker* fantasy), Ken decided he would use this outdoor furniture as his office and, having decided upon it, inspiration struck him. Or rather, various very odd people approached him for a chat and he began scribbling down what they told him. It was here that he wrote, or claimed to have written, *Furtive Nudist*, the first part of what would become the *Bald Trilogy* of monologues. Much of it was set here, too. The trilogy is his masterwork, the crowning achievement of his long, odd, mind-bending career. Across the three plays—which begin with Ken's surreptitious streak across the sand dunes near his dad's retirement home in Cornwall and end, finally, with an ecstatic vision of Prince Philip as Godhead while Ken trips out of his head high up a volcano on one of the South Pacific islands—Campbell weaves together memoir, philosophy and outrageous lies into a truly unique, reality-destabilising machine: 'I'd seen the Truth but in the same instant corrupted it.' Reviews of the *Bald Trilogy* make liberal use of the word 'genius', but then, as Ken pointed out, 'They call me a genius so they don't have to give me a job.'

To understand fully the significance of Ken's *Bald Trilogy*—and in particular its creation myth—we first have to travel back another twenty years to 1970. At the time, our hero was running the Ken Campbell Roadshow, the first venture in which he took complete control, a company who performed urban myths, shaggy-dog stories, dirty jokes and, later, stunts and feats of escapology, around the pubs, shopping centres and working men's clubs of northern England. In support of the Roadshow's work, Campbell wrote an article for a local arts newspaper in Liverpool in which he criticised Arts Council funding for the theatre: not that there should be more of it but that there was any at all. Ken's complaint wasn't a proto-Thatcherite call for the purity of the free market, however, but a plea for less respectability in the acting profession. 'Instead of being the safe, subsidised, castrated toys of the middle class and the bourgeoisie, actors, once again, would be the outcasts, the outrageous, the rogues, the vagabonds, the gypsies. What I mean is—they would once again have dignity.' Compare that to a quote from Robert Pogue Harrison, in the book I mentioned in the last chapter, *Forests: The Shadow of Civilization*. Harrison explains how, in early medieval times, the introduction of Forest Law had created a hiding place for more than just deer: 'In them lived the outcasts, the mad, the lovers, brigands, hermits, saints, lepers, the *maquis*, fugitives, misfits, the persecuted, the wild men. Where else would they go?'

That the forest is a key site of discombobulation is enshrined in its very definition: the meaning of 'forest' has nothing to do with trees. It is, strictly speaking, the king's hunting land, governed by Forest Law rather than the usual common law. All the wild animals of that land are the king's property, and the land has to be maintained to the benefit of those animals. That means no building of fences, no farming and no removal of trees or other greenery without the permission of the king or his designated underlings. *Foris* is Latin for 'outside', literally outside the standard law but also outside of any enclosure. Even more obscure, the word *forestare* means 'to keep out, to place off limits, to exclude'. To *afforest* isn't to plant trees; it's to declare, as sovereign, that the land is reserved for your recreational purposes. As a result, much as there are at least two versions of me in the forest—the me telling you all this and the me acting wet next to a pond— there are at least two forests as well.

First, there is Epping Forest as it exists today, which was bought for the nation by Act of Parliament in 1878. This preserved wilderness starts at Forest Gate in east London, seven miles east-north-east of Mansion House, a scrubby area that can't decide whether it's inner city or suburb, only a mile or so north of the traditional home of West Ham United. The forest then runs in a narrow ribbon up through Wanstead, Leytonstone, Walthamstow and Highams Park until reaching Chingford (ten miles north-north-east of Mansion House) and opening out into the broad swathe of woodland that flows for six miles or so out into Essex, past Buckhurst Hill, Loughton, then Theydon Bois and up to Epping itself, terminating—the dot on the i—at Wintry Wood, just north of that small market town. It's these six thousand acres which form the core or heart of this book. They make the shape of a bow aimed at the heart of London, fourteen miles long. One of its thinnest strips, between Walthamstow and Chingford, was where I first explored when I moved to the area fifteen years ago—little more than a path between houses in winter when the leaves are gone, but a space that seems to expand in summer in direct proportion with the growth of foliage and loss of sight lines. And this expansion hints at—remembers?—the second, phantom forest, the great Forest of Essex, declared as hunting land for the use of the king by William I (aka William the Bastard) after his coronation in 1066.

Here we hit upon the national crisis which formed the forest. Having defeated Harold at Hastings, William and his band of Normans swept to power, carved up the land amongst themselves and, in the process, disenfranchised the Saxon nobility. Yes, the new king was a keen huntsman but he was even keener to rub the noses of the Saxon population in the fact of his victory. Afforestation was an excellent way to do this, to assert his rights as monarch, and he took to it with such brio that the *Peterborough Chronicle* (something of a Saxon moan-fest) claimed 'he loved the stags as much / as if he were their father', which implied he loved their mothers even more. In protection of these rights, the law stated that an illegal stag-killer should have his eyes plucked out. There were regulations about what type of dog you could keep and how its claws should be clipped. A whole set of rules governed what you could do to bushes and when. A separate and

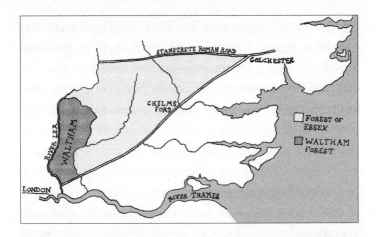

parallel court system enforced these edicts, an ominous and slow process which could catch you up in its functionings for years before spitting you out at the other end. Among the many gigantic areas that William set aside for his leisure pursuits was most of the county of Essex, stretching out east and north of London.

It's in the nature of kingship to repeat a good trick with no one quite brave enough to say when it's getting tedious. The monarchs who followed William extended his afforestation until finally, with Magna Carta, which devoted four clauses to it, they were made to stop. Under further baronial pressure, Henry III firmed all this up in the Charter of the Forest, actually ranked in importance alongside Magna Carta at the time, or even considered part of it. In Essex, over the next fifty years or so, knights were sent on a series of perambulations to establish exact borders and shrink the boundaries back a little from the greatest monarchical extravagances. What was left was a large triangle of land with Stratford (now in east London and best known as the site of the 2012 Olympics, just under four miles north-east of Mansion House) at its southernmost tip. The north-western corner of this forest would eventually become Stansted Airport, twenty-six miles from Stratford, a Norman Foster-designed hangar from which millions of Brits take cheap flights to Europe and then return, sometimes sunburnt, often hung over, almost always ungrateful. The eastern point was provided by Colchester, thirty-three miles east of Stansted, not that far from the sea. The overall area of the Forest of Essex was well over four hundred square miles, or two hundred and seventy-five thousand acres.

Waltham Forest, as it became known during the thirteenth century, was the south-western corner of this expanse, a highly irregular rectangle with its base formed by the road between Stratford and Romford and running north thirteen or more miles. Its area was a mere eighty thousand acres. Most of the southern section of this land is now a vast, low-rise suburb, one part indistinguishable from another, its avenues built from tectonic concrete sections, the houses made small by too much sky. The north is a cellular structure of roads and fields. The bow of Epping Forest is contained within its boundaries, like a wisp of smoke rising up a rudimentary chimney:

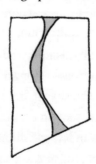

The location of Ken Campbell's pondside composition of the *Bald Trilogy* was important to me because the Hackney side of the River Lea—where his flat was—lies in the old county of Middlesex, while the eastern side, where the pond is, lies in Essex, inside the old boundaries of Waltham Forest. Campbell's crisis had led him to the forest, where 'the outcasts, the outrageous, the rogues, the vagabonds' all go to hide. Or rather, it led him *back to* the forest. Ken was born and brought up in Gants Hill, a fairly nondescript suburb just north of Ilford and very much within the borders of the historic forest, too. And it was in Gants Hill, specifically at Gants Hill Library (three and a half miles east of my house in Walthamstow, eight and a half miles north-east of Mansion House), down in its basement, that Ken claimed, in the third part of his trilogy, *Jamais Vu*, to have been party to an earth-shattering revelation.

This was early in Ken's career as an actor, during one of the crises of unemployment which afflict all but the luckiest thespians. He had returned from a stint in Bournemouth working on one of their swimming pool-based aqua-shows, and now he couldn't get another job, the aqua-show skill set not being hugely transferable. He began going to the library and, in particular, down to the basement, where all the books that no one wanted to read were hidden away. There he discovered a group of scholars intent on learning all that everyone else considered to be excess to their requirements. This gang had adopted a simple but profoundly liberating philosophy, one which Ken would adhere to for the rest of his life and which is symptomatic of the forest way of thinking. 'Don't *believe* in anything,' Ken claims these savants taught him. 'But you can suppose *everything*: and in fact you should.'

I was looking for revelation too. I had already developed a theory that Margaret Epstein had cast some sort of spell on Jacob to drive him mad out in the woods. The only problem was that I had no evidence, not a shred; not, hand on heart, even a hint. There were simpler readings involving lust, jealousy, codependence, revenge and so on, the usual ingredients of a marriage. But then again, I thought there was something to it, felt it when I was out there under the trees, wandering around on my own, with wood pigeons exploding from the branches above me and the noise of silence welling up round me. It was more than just Grimm or Harry Potter or all those other literary references I fired off to buttress my hunch and make it sound clever. It was *out there*, it was *real*. It was inside me too, a kind of hunger that was satisfied when surrounded by beeches. I was looking for magic, looking for it everywhere, for some miraculous leap beyond my own existence, for something I didn't believe in. In many ways, the made-up book I wasn't writing—and which I didn't believe in either—*was* that magic, and so the trees had to be too. I wanted to look down between my feet, see a wand lying there, pick it up and wave it and for everything to be done.

There are two types of writer: those who are scared of magic and those who think they've mastered it. Like all kinds of people, writers tend to have very little money and very little control over their fate. Like very few people except for the unemployed—to whom they bear a marked formal resemblance—they have an awful lot of time to ponder this. The uncertainty, the reliance on the whims of distant and unpredictable gods (agents, editors and, if you're lucky, readers), is ideal soil for cultivating superstition. A day spent sitting at a desk failing to achieve anything acts as sunlight for the mutant plants that grow, so that they writhe up strong from the insecurity, a forest made of ladders under which the writer finds him or herself walking. Most of us scuttle between these

tall trees expecting a demon to jump out and eviscerate us—
we live in fear. The others, the chosen few, the Grand Wizards
(always wizards), end up thinking they've conquered language itself, that they've tamed words, and hence that they
own the forest and can ascend those ladders to heaven. Ken
refused to belong to either group. Wizards are boring, far too
keen on throwing their weight around, the kind of people
who corner you in a pub and tell you all about their *vision*.
And the rest of us, the quaking (m)asses? We're pretty tedious
too, hopping from foot to foot as we throw spilt salt over both
shoulders, unsure which of them the devil lurks behind. Much
of Ken's working life was devoted not to harnessing magic
nor to guarding against it, but to setting it free. His forest
formulation—that you should believe nothing but suppose
everything—was a kind of spell, a liberation from narrow
rationalism, and, pretending I had the guts to do it, I decided
to use it as such.

Sitting in an actual valley as opposed to halfway up a hill,
to the south of the mystery of Court Hill, Baldwins Pond is
a brown, undefined depression of open water girded on one
side by a popular track that, for well over six months of the
year, is little more than a mudslide. If you can make it round
and past this path—and much of your time is spent *avoiding*
it rather than following it, picking through brambles searching for somewhere more solid—you come to a steep, open
incline on the other side, a grassy hill punctuated by a single
sycamore with a bench beneath it and crowned by a row of
houses at the top. There's always a line of cars and vans parked
on the road, a good proportion of them left there by the professional dog walkers heading downhill, their hands bursting

with leads like alien spaceships scanning the terrain for intelligence. Once you've slogged your way up the slope you can either keep walking straight into the pub opposite or take a left and find Deerhurst, the second property the Epsteins rented. Just before it, though, is a strange interloper, a wooden chalet allegedly imported ready-built, dragged by horses all the way from Switzerland sometime in the nineteenth century. This is where Ken Campbell lived his last few years, retreating here from Hackney after one of his dogs (not Werner) bit a policeman, only to be saved from being put down by taking a fictional starring role in a non-existent dog display team.

The house was for sale. Perhaps if I had £700,000 to spare I would have bought it. It's a fantastic location if you're thinking of West Essex—and somebody must be. I examined the pictures online, all airbrushed and photoshopped and generally gussied up until they looked computer-generated. The owners had added their own comments to the online spec for the property:

> When we bought this house 3 years ago it was in a terrible state, but we loved the location, the fact that it was nestled away in the forest and it had a great feel about it. Since living here we have completely remodelled the house and, as well as restoring the listed part of the house, we also added a large ground floor extension to make a large kitchen/diner. We love entertaining and the open plan layout of the house has enabled us to do that with ease. We feel the loft is a great selling point as it is a very usable space, it is as big as the main bedroom, has natural light and oak flooring throughout.

Somehow, when I read that, Ken felt very dead indeed. I decided to consult a psychic.

This was out of character. I am not the psychic sort. I'd go as far as to say I am the opposite of the psychic sort. My habitual attitude to anything even vaguely *magical* is that you deny its existence even if you've already been turned into a frog—and I maintained this stance unwaveringly while questing for the very thing I'd rejected. But I fastened onto my psychic scheme like a shipwreck survivor far out at sea with only a piece of wood for company. The notion was inspired by videos I'd seen of Ken on the internet—inside this very house, the third point in his forest triangle—talking about seances. There was a very funny one where he was advised by a dead ex-girlfriend to put up his bathroom cabinet, and the lady sitting next to him was given help from her dead husband on the use of venetian blinds. There was another in which the ghost of Laurence Olivier told Ken that the world's greatest living actor was Jackie Chan. I'd also read a story from 2004, around the time Ken moved to Loughton, of the renowned 'psychic' Derek Acorah and his TV crew getting lost in the forest at night while communicating with Dick Turpin's ghost. Apparently he told them he enjoyed riding around but, despite over two hundred and fifty years of local knowledge, couldn't point them back to a road, so that they had to phone for a Forest Keeper—the people who police these woods—to come and rescue them. So, yes, I suppose the starting point was that I thought it might be funny, and hence in keeping with Ken's project. After all, when asked about humour in his epic production *The Warp*, he had replied, 'I don't think anything which isn't funny should ever be put on. I think serious drama is a very dangerous travesty of life.'

There's a theory that to write anything that matters, anything worthwhile, you have to confront what you're frightened of. It's one of those artistic clichés I'd been wrestling with ever since I was a teenager, flopping in its grip like Jacob in the arms of the angel. The thought of going to see a psychic, I discovered almost as soon as the idea occurred to me, terrified me. First, I was repulsed by the dishonesty of paying someone to do something I didn't believe in and I was afraid that I would give myself away. Second, I was nervous about going to a strange house to meet someone I didn't know to pay them money to do something completely disreputable. Third, and most importantly, I was afraid it might be *true*. Because let's be honest, why would a psychic even think I didn't believe them unless *they were actually psychic*? And as long as I was paying, why would they care? If I feared tricking an estate agent into showing me round a house I had no intention of buying—and that proved to be a bridge too far for my delicate sensibilities—imagine how I'd feel about letting someone know, however accidentally, that I considered them at best deluded and at worst a crook.

Underlying this scepticism, though, was the terrible hope that it would indeed be true. I imagined to myself that I was pretending I wanted Ken to be Virgil to my Dante, but at the same time some part of me thought that maybe he would. How much easier things would be if I had a ghost directing me! That I had found no better way to navigate through life was shown clearly by the predicament I found myself in. Plus, the fact that I didn't want to do it, was afraid to do it, showed me that I really should. I'm not trying to claim that any of this makes sense. Not now. Though at the time it did.

Most important of all was that first axiom of forest thinking: 'Don't *believe* in anything. But you can suppose *everything*: and in fact you should.' Nothing is to be discounted as nonsense, but nothing is to be embraced as truth either. Everything is to be held in a delicate balance, non-binary, mutable. Later I came to believe that Ken had taken the idea from Robert Anton Wilson, and that the autobiographical integrity of the *Bald Trilogy* perhaps left something to be desired, although in terms of the point he was making I suppose that was the point. And the point remains exhilarating. The reality, however—at least in relation to my psychic episode—was a smidgen less compelling.

It was bound to be, considering the contradictory impulses I was trying to accommodate. I wanted it to be funny, but I wanted it to be revelatory too. I wanted it to be ridiculous and I wanted it to be brilliant. I wanted funny voices, drama and mayhem, cryptic clues and admonitions and cloaked instructions, but that wasn't how it turned out. I picked a psychic near to Ken's old house in Loughton—in case he was hanging around, I suppose. I thought she'd at least close the curtains and burn some incense, try to create a little *atmosphere*, but that wasn't the way she did things. We sat there one bright morning, our phones on the table, glasses of water shaped like Coca-Cola tins, a cup of herbal tea, a pack of divining cards which looked like they were bought off the internet instead of passed on to her by a withered crone. Later she laid them out on the table in front of me and the pictures on them were so saccharine, so self-evidently *awful*, that I squirmed and wanted to cry even as she told me of the manifold triumphs that awaited me. There were a couple of books on esoteric forms of healing propped against the carriage clock and a North American Indian dream catcher of dubious provenance pinned up on the wall, but it wasn't exactly a witch's den. Even the medium herself looked unremarkable—tall, rather frumpily dressed for someone in their late twenties or early thirties, with a loose handshake and a matter-of-factness that was probably meant to be reassuring. I didn't want to be reassured, I wanted to be terrified. When she told me that she didn't vomit green ectoplasm it was all I could do to hide my diappointment.

The medium seemed completely sincere to me. We started off with a pain in her left elbow or arm which, she said, might be a sign of rheumatism or left-handedness in Ken. She saw a tennis court and thought of tennis elbow. She mentioned Karl Marx, which made me curious, then went into a riff about cigars and eyebrows until I realised she meant Groucho. She worked on a free-associative basis, telling me whatever came into her head and hoping that something piqued my interest or tallied with what I wanted to hear. There was a problem, though, in that she didn't seem to know what I wanted to hear. Which is probably fair enough as I didn't know what I wanted to hear either. On she went, on and on. On and on and on. *An old-fashioned record player, surgery, stomach cut open, Baldwins Hill, a small white dog, a sedate wife with her hair in a bun, a well-kept garden, Epping, the town of Epping, Theydon, railways, the old-fashioned kind, not the tube, someone far out, Martin. Who is Martin around you?*

Martin is no one around me. It kept coming and coming. I found my attention beginning to drift a little, saying yes, uh-huh, yes, to whatever she came out with. I was too embarrassed, too English, to say no, I'm sorry, that doesn't mean anything to me. Which is ironic, because part of the reason I was there was to find a way to short-circuit my urge to agree, my need to be liked. I couldn't stop checking the time on the carriage clock, wondering how much longer I had to go before I could escape. Did I sense the presence of Ken? No, I did not. The Ken Campbell she believed herself to be communicating with was a gentle, chuckling, warm-hearted old fella, inordinately pleased that I should be interested in him and very keen to reassure me that everything was going to work out for the best. Which didn't sound like Ken at all. Where was the mischief, where was the mayhem? Where was Ken's legendary temper? Where was his unreasonableness? This is the man who ordered an actor to attend every single Ken Dodd performance, matinees and evenings, anywhere in the world, for the next year. Who, when asked by another actor for help with a scene, told him to imagine a ball of white-hot molten metal in his hand. The actor closed his eyes, clenched his fist, thought for a while and said he could feel it. 'Good,' Ken replied. 'Now shove it up your arse.'

It's amazing how much someone can actually *say* in half an hour if they keep talking and talking and the other person doesn't interrupt them or offer any help with the conversation. It wasn't even as if she spoke very fast, just slowly, calmly plodding through whatever she was *getting*. If there was a theme to be extracted from what she told me it's that I am destined if not for great things, then at least to get this book published. The beauty of it as a prediction is that if you are reading this, it turns out she was right. Shazam! It was the instrumentalism that irritated me the most. The psychic said it made a change not to be discussing someone's love life, but she couldn't quite jump loose of the idea that I was there to be told something about myself and my future. When in reality I was already heartily sick of myself and was hoping to be entertained or educated by a dead man. Preferably both.

The comedian Stewart Lee has described Ken as a 'shaman-clown and Zen-ventriloquist', which starts to hint at the man's qualities. Equally enlightening is a comparison drawn by the actor/director and now Thomas Cromwell TV crumpet, Mark Rylance. Towards the end of his life, Ken started a company that specialised in improv with a Shakespearean flavour, the School of Night. Rylance, then creative director at the Globe Theatre, had them in to do a performance but regretted, later, not getting Ken back for more. He compared him to the first great fool in Shakespeare's theatrical company, Will Kemp, allegedly pushed out because he annoyed the Bard of Avon by improvising and mucking around too much. Shakespeare's fools have been described as *reality instructors*, a phrase taken from Saul Bellow, in whose books a variety of lawyers, hoods and rogues educate his central characters in matters of perspective. It's the perfect description of Campbell's methods and aims, that combination of humour and hitting you over the head with a big stick, though his *realism*, both theatrical and metaphysical, was liberatory, expansive—explosive, even—rather than being designed to rein anyone in. The main lesson Ken wanted to teach is that reality is much odder, much more unstable, much funnier, much less *safe* than we might believe. He's not trying to bring anyone back down to earth. That would be irresponsible.

In pursuit of this aim, Ken employed different techniques at different times in his career, most of them involving a joke and a cudgel. In the late 1970s, he and his new company, the Science Fiction Theatre of Liverpool, staged Robert Anton Wilson's *Illuminatus! Trilogy* to great acclaim. The novels, dealing with an age-old clash between the Illuminati (evil men in suits) and the Discordians (hippie worshippers of chaos), aimed to repurpose paranoia and conspiracy theory for an emancipatory cause, perhaps not entirely successfully. Wilson himself came over from America to see the play when it transferred to the brand new Cottesloe auditorium at the National Theatre, where he supplied Bill Nighy with performance-*enhancing* LSD and declared, 'I've come to the conclusion that this isn't literature. It's too late in the day for literature. This is magic!' That was certainly Ken's aim, though I'm not sure he was there yet.

As envisaged by Campbell, the *Illuminatus!* play was broken down into five sub-plays. Each sub-play had five acts and each act was twenty-three minutes long (the number 5 is very important to Discordians, as is 23, because 2 + 3 = 5). That meant the entire play lasted for about ten hours. The original rationale for the play's length was presumably the amount of material needing to be covered (the trilogy is made up of three very long books) plus a certain relish for annoying people. I believe, though, that Ken discovered that something happens when you produce a play that long, in particular that, at some point very close to the end of his epic, he noticed a *change*, an odd ripple running through the audience that he needed to explore further. That, at least, is my explanation for his next big production, *The Warp*.

Based on the life of its writer, British counterculturalist Neil Oram, *The Warp* is in the Guinness Book of Records as the longest play ever, clocking in at somewhere around twenty-three hours. But the whole purpose of making it so long was that at some point after the ten-hour mark it ceased to be a *play* at all. Andrei Tarkovsky, the Russian film director—an unlikely ally for Campbell, despite the *Stalker* fantasies, in that he seems to have taken himself and his work deathly seriously—was a fan of the cinematic long shot, indeed, the very long shot. Discussing it, he explained that 'if the regular length of a shot is increased, one becomes bored, but if you keep on making it longer, it piques your interest, and if you make it even longer, a new quality emerges, a special intensity of attention'. What did *The Warp* become? What new quality emerged? Something beyond *performance*, perhaps, something which, not least by dint of how long it had stayed in the audience's life, became part of their reality. As one of them commented, staggering out, wide-eyed, smiling beatifically: 'I've just had one of the most amazing experiences I ever had in the theatre. It's absolutely mind-blowing. What Ken Campbell is doing is transforming theatre. It's transforming the audience.' The performance becomes ritual, the ritual becomes reality, reality is changed. Magic.

Mired in disappointment, which soaked into me like cold marsh water, one phrase from my consultation continued to itch away at me: 'a sedate wife, a calm lady'. The medium herself had been dressed as 'a sedate wife' for my visit, even though on her website she presented an image which was much more *mystical* (yes, I mean shawls). I wonder whether this drag was adopted because, inferring from my emails about my 'historical project', she had expected me to be older and more respectable than I turned out to be. But thinking about how the medium had adopted a disguise for my visit reminded me, in turn, of Sophie Firebright.

In the second of his *Bald Trilogy* plays, *Pigspurt*, Campbell expounds at some length upon his love of the later works of Philip K. Dick, the Californian science fiction writer best known for the vertiginous, reality-morphing stories which inspired *Blade Runner* and countless other Hollywood movies. Less often touched upon is Dick's religious-psychotic conversion/episode(s), which was Ken's particular field of interest. Although *VALIS* was the book he tried to adapt for the stage, in *Pigspurt* Ken focuses on a word he came across in Dick's *Exegesis*, 'enantiodromia' (intriguingly, *Exegesis* wasn't published in full until 2011, whereas *Pigspurt* was first performed in 1992). As if letting us in on a secret of the Ancients, our monologist claims to be only the fourth person to have used this word, after Heraclitus, Jung and Dick himself. Enantiodromia is 'the sudden transformation into an opposite form or tendency' and Campbell used it, he says, to develop workshops on the Enantiodromic Approach to Drama (he was, he admits, already doing the workshops before he had the word, but it sounded more impressive with it and hence he could charge a higher fee).

The basic theory of the workshops, Ken explains in *Pigspurt*, is that 'a face is comprised of two contradictory, mutually exclusive *facets*'. Ken's facets are, on his left side, the Spanking Squire ('he's into the chastisement of the young ladies of the village whether they've been naughty or not') and, on his right, 'a vacant, inept housewife called Elsie'. During the course of what follows, he happens across a shop round the back of Euston selling gear for transvestites and ends up transforming himself—and, by extension, the hapless *Elsie* side of his personality—into Sophie Firebright, his own version of Dick's St Sophia, the Gnostic Goddess of Holy Wisdom. It seemed reasonable to me then (and I use the word *reasonable* advisedly), to suggest that my medium had put me in contact with Sophie Firebright, 'a sedate wife, a calm lady', the goddess of wisdom. What would Squire Pigspurt want with me, anyway? Perhaps Sophie was pointing to my own role as a vacant, inept househusband when s/he tried to put an apron on me. Perhaps I had my own Elsie. That was where it started, this idea of myself as both the actor in a play and the director of that play, of omniscient narrator and comedy chump, my dual nature, double-faceted.

Once I knew what I was looking for, Ken even offered guidance as to what sort of actor I should direct myself to be. There's documentary footage of the Ken Campbell Roadshow, including an interview with its young boss. Back then he had long, fine hair and high, delicate cheekbones. He looks rather beautiful, before age and tactics have forced him to rethink himself as a gargoyle, smoothing his hair back from his face, that slight smile already there, a calm mischief, the eyebrows wide but controlled, two sleeping question marks. Explaining how you deal with a crowd outside of a traditional theatre setting, Ken says, 'People get upset if they think there's a script. I don't know why. It puts people's backs up. So it should be as if unscripted. Or if it looks scripted then we all know it is, don't we? So there's a sense in which a lot of it's consciously *bad* acting.'

This obsession with bad acting, or what he came to call *gastromantic* acting (in that it involved projecting from the arse rather than the stomach) was one of Ken's key concerns and, though he dressed it up as a joke, it goes some way to illuminating not just his attitude towards the theatre or to art or even to life, but to reality itself. To fall foul of naturalism is to buttress the status quo, it is to soothe the audience, to make it easy for them, it is to be their safe, subsidised toy. There is, though, no status quo which can be maintained, only an infinite forest of paradigms, each of which can be supposed but should never be committed to. The gastromantic takes Brechtian alienation to a whole new level—to a whole new universe, you could say.

N arrator-I, desperate for a good story, told actor-me what to do. I had recorded my visit to the spirit medium, so I took the transcript I'd typed up and cut out everything she could've guessed or dropped in because it was an easy thing to say. I excised her attempts to reassure me or convince me I had a glittering literary future. I cut and cut until the two sheets were fine, filigree things, snowflakes from the aftermath of some future data apocalypse. I chopped up what was left, took the individual words and phrases that survived and put them in a sun hat, on the grounds that a sun hat is always welcome to a bald man. I took the hat, got in the car (it had been raining and I was in a rush and even walking writers occasionally need to get somewhere fast) and went to the pond. Solemnly, attempting a gastromantic delivery despite the embarrassment, I made offerings to Ken—a small pebble from Pembrokeshire that I was rather attached to and a slug of good whisky from a hip flask for libations. Then, one by one, I picked the phrases out of the hat and stuck them down with Pritt Stick onto a fresh sheet of A4, the whole balanced in my lap, an uncomfortable, messy ritual. This, I thought, would be my map, my guide through the forest. As I looked up from my work, the sun raked out from behind a cloud and a rainbow formed above the railway tracks. This was my way out. Ken would lead me through the maze to the exit.

The problem was, of course, that what I was left with made no sense. Magic's like that. If it made sense it would be science.

CHAPTER 3

The Doom Tree | The Lovely Illuminati Tree

I have never been a very successful tree-climber. For one thing, I lack strength in my hands and wrists and arms. For another I lack conviction. I am more scared of falling than I am excited by the prospect of succeeding. You need to commit if you want to get up a tree. You can't really do it half-heartedly. If I was being kind about myself I'd say that to do it well you need a lack of imagination—it acts as a dampener on your spirits if you're forever picturing your twisted, broken body, blood seeping from your ears, after the fall. On the other hand, it's probably more accurate to say you need a different type of imagination, a sense of curiosity, or *adventure* if you prefer, to focus on how the world might look if only you could get a little higher. All the same, I started searching for trees to climb. To the extent that I could be said to have a job description any more, this felt like it must be part of it. I indulged reveries in which actor-me exceeded director-I's expectations and swung from high branches wrapped in animal skins, a wild man of the trees, an east London orangutan only better-looking and more dashing. It remained fantasy, my orders to climb higher

61

and harder met by myself with a certain degree of truculence. When I tried it first I'd done it in the hope of a view, but I'd learnt the folly of that approach. Instead I settled for finding somewhere to nest, this big, uncomfortable, flightless bird, keeping warm the egg of my head.

It turned out it wasn't as easy as it might sound. It's ridiculous, but in a forest of thousands upon thousands of trees, very few of them struck me as ascendable. The birches were too thin, the hornbeam too *waxy*, somehow. Oak looked like a good bet until you got close enough to realise just how far above you the lowest branch was. I had no tricks, no ways to shinny up a trunk—I needed something low enough to pull on. They would always have a doughty beam jutting proudly out inches above the reach of my weak and flapping hands. Only the beeches offered any hope, and only then because of the interventions of previous forest-dwellers. On some of them the grown-out pollard of centuries past provided a platform on which, if I could haul myself skyward, I thought I could crouch or sit. Pollarding or lopping—cutting back branches so that fresh timber can grow—was traditionally carried out just above the height that deer or cattle could reach, so that the new shoots would develop without being eaten. Usually a little way above the top of my head, it was difficult to get up to, yes, but not impossible. Every time I did it, I would reach my target breathing hard and shaking all over, conscious that I might topple out the other side. That's the thing with trees. There is always another side to the one you ascended and it's always much nearer than you hoped.

I was increasingly enamoured of the *enantiodromic* approach I'd picked up from Ken. The side for climbing up a tree and the side for falling down. The meek and fearful aspect of my face

and the cruel and all-seeing. The descent into crisis and the—
so far putative—ascent back out. The forest as threat and the
forest as refuge. The loving and vengeful facets of both the
primitive goddess and the artist's wife. The artist himself as
manipulator of mythic symbols and facile modeller in clay.
The Shakespearean fool as joker and instructor. The audience
as inhibitor and stimulator of sexual intercourse. The winding path to enlightenment and the winding path to nowhere.
Magic and science.

I had never heard of Mary Wroth before I started mucking
about in the forest. I first came across her in Sir William Addison's 1945 book about the forest, where she was treated as an
adjunct to the playwright Ben Jonson. Until the 1960s no one
was particularly interested in her. Since then her reputation
has grown, but only really in academic and feminist circles.
Nevertheless, she is a pioneer. Her book, *The Countess of Montgomery's Urania*, was printed in 1621, making her the first woman
to publish an original full-length prose work in English. In
addition, with *Pamphilia to Amphilanthus* she became the first
Englishwoman to write a full Petrarchan sonnet sequence.
Her play, *Love's Victory*, is also held to be the first original play
written by an Englishwoman, although as it was never performed or published during her lifetime it's hard to be sure
how that particular record is measured. And while records
are all very well—even ones that make no sense—it's worth
putting what this might have meant into some sort of context.

The Countess of Montgomery's Urania was an early modern
romance at a time when there was a moral panic about
women even *reading* romances, which were thought to excite
the passions in a way well-born ladies would be unable to
control. To compound the outrage, many readers of the book

found reference in it to contemporary characters and events. In particular, Lord Edward Denny, Baron of Waltham—who lived only a few miles away from Wroth's home at Loughton Hall—was horrified to discover a character, Sirelius, whose father-in-law attacked his own daughter for committing adultery: 'Her father, a fantastical thing, vain as courtiers, rash as madmen, and ignorant as women, would needs (out of folly, ill nature, and waywardness which he called care of his honour…) kill his daughter.' There was a rumour going around that Baron Denny had behaved in exactly such a manner when his only child, Honoria, was accused of cheating on her husband. You would think in the circumstances it would be best to ignore the reference rather than drawing attention to it, but there is no reason to believe Denny to be a man of any great intelligence or sensitivity. He once tried to evict a family from their home by surrounding their residence with twenty armed men and starving them out. Instead of ignoring the perceived jibe, he wrote a poem called 'To Pamphilia from the father-in-law of Seralius' and sent it to Wroth as well as allowing others to see it. In it, he called Wroth a 'Hermaphrodite in show, in deed a monster', finally telling her to 'leave idle bookes alone / For wise and worthyer women have written none'. In his covering letter he drove home this point, suggesting that she 'redeem the time [wasted] with writing as large a volume of heavenly lays and holy love as you have of lascivious tales and amorous toys'.

Wroth responded with an apologetic letter in which she claimed that she had never meant the character of Seralius to be thought of as Lord Denny's son-in-law, Lord Hay, and that in any case, she had never intended for *Urania* to be published. She was on dangerous ground. As a baron, Denny was of a higher social status than she, and Lord Hay was one of King James's favourites, a fellow Scot who owed his position in England to the crown. But having struck a regretful tone, Mary enclosed a poem replying to Denny's line for line, starting 'Hermaphrodite in sense, in Art a monster' and finishing 'Take this then now lett railing rimes alone / For wise and worthier men have written none'. There's an instructive contrast between the slipperiness of the apology and the counterpunching of the poem. Wroth plays two roles in one envelope—refined, subservient gentlewoman and rogueish Jacobean writer—and carries them off effortlessly. She is another double-faceted forest-dweller.

Urania is epic in scale. The first volume, as published in 1621, ran to only 280 pages, but the second volume, found among papers at her family's home and recently published, is much longer. The whole enterprise adds up to some six hundred thousand words—six times longer than the book you are holding—and features over three hundred characters. In theory, it tells the story of the on-off love affair between the cousins Amphilanthus, King of Naples, and Pamphilia. But this basic plot (or non-plot, as it lacks not only an end but also a middle and even a beginning) is then refracted and echoed in the stories of hundreds of other noble folk, who travel around the world being deposed from their kingdoms or recapturing their kingdoms, falling in and out of love, trusting and not trusting their lovers, disguising themselves or being misled as to who they are, having challenges set and often failing them. Even abridged and with modernised spelling I found it impossible to make my way through. The effect is a little like trying to follow the opening plot of every Shakespearean comedy all at once, with the same actors taking multiple parts and switching costumes. It's less like a story than a pathology. The only thing I've ever read like it is the work of Raymond Roussel, a distant inspiration for my own failed psychic word game out next to the pond.

Roussel, a kind of anti-Proust, generated his texts using homophones. He would think of a sentence and then come up with another sentence that sounded almost exactly the same but which had a wildly different meaning. The purpose of the writing was to find a story which bridged the gap between the two sentences. In later works this homophonic punishment was layered up until it, and the stories, almost ran out of control. The results are strange and ornate and unsettling and often unreadable. He was a monumental failure in his own lifetime, later adopted as a pioneer and forerunner by the Surrealists and then the Oulipians, the France-based champions of formal constraint founded by the writer and mathematician Raymond Queneau. Unlike Roussel's, though, Wroth's pathological prose doesn't appear to be generated by the text itself. Nor is it as voluptuously odd, or surprising. Wroth claimed that she never meant the work to be published, that it was written only to entertain family and friends. Lucky them. The sheer repetitiveness of its themes, the constant succession of crossed loves and deposed monarchs, of wronged women, trafficked women, trapped women, makes for an unlikely entertainment. It's boring and it's confusing. I found myself thinking that there had to be something else going on, some way to decode it.

There are two beeches out in the forest which look definitively old enough to have been there when Mary Wroth was living a short ride away at Loughton Hall. If you plot every last place that I talk about in this book, these trees sit pretty much in the middle, about a mile and a quarter due west of Baldwins Hill, in an area of the forest called Hill Wood. In the mid-1940s, Jacob Epstein used to wander up from Loughton with his son Jackie and take him to tea at the Robin Hood pub. It's still there, though now specialising in Thai cuisine. If you carry on west from there, past the tea hut where bikers gather at weekends—the first speedway track in Britain was built further up the hill and they come here in honour of it—you're in with a fair shot of hitting these trees.

I call the larger of the two the Doom Tree. It's a monstrous thing, a chthonic deity dragged into sunlight. It must have a diameter coming on for three metres, and every available space has been carved and re-carved so that from a distance the bark seems to belong to a whole new species. One long, muscular tentacle of a branch has 'YE ROYAL BATTLE FORT 1968' tattooed along its underside. There are hearts and anarchy signs and initials, an 'LSD' and dates—1960, 78, 87, 72, 08, 09, 2000, 56, 1947. Another branch, which you can only read once you clamber up, says 'INVASION IMMINENT'. There is also a fine little bird, standing, wings folded, pecking at someone's initials, with the number 3656 written under its delicate feet. But my favourite graffito, lovingly done in boxy capitals, with long arrows leading to it from every direction, the nearest we'll come to a wooden early modern sonnet, reads 'THOU, WHO TREADETH ON THIS TREE, THOU, SHALL BE DOOMED'. It's high enough up that you're already doomed by the time you find it, its effect mitigated somewhat by the pithy reply scratched in underneath: 'BOLLOX YOU CUNTS'.

•

In the traditional view, Mary Wroth's writing grew out of the series of crises or cataclysms that defined her adult life, crises that she wore like insults scratched into the bark of her psyche. The first crisis was to be a woman, and many of the others flowed from this accident of birth. But they also took in most of the schisms and insecurities of her era, which ran from the late reign of Elizabeth I (who she danced for as a young girl) to the Civil War, a period of unprecedented change in the social and intellectual order. It's also when magic and science definitively broke apart, becoming not merely separate disciplines but antithetical world views.

Mary Sidney, known to her father as 'little Mal', was married, aged seventeen, to Sir Robert Wroth, a keen huntsman with estates at Loughton and Enfield. Depending on which book you read, he was either a drunk and a gambler or a Puritan (plausibly both). Either way it was hardly a perfect match for someone who had read her uncle Sir Philip Sidney's *Arcadia* aged ten and most volumes of Spenser's *Faerie Queene* before they were published, and who had been renowned as a child for her sharpness and intelligence. Her father was, however, paying the price for his closeness to Robert Devereux, 2nd Earl of Essex, who had been beheaded in 1601 for leading perhaps the lamest coup in the history of angry men riding around on horses and mouthing off a little too much. He was also very short of money. Unsurprisingly, the wedding wasn't a great success, possibly due to sexual problems or Dad's failure to pay the dowry. But the Wroths had money and the Sidneys still had some good contacts at court, so everyone seems to have bumped along. Mary played a full part in court society and, perhaps more notably on a personal level, literary society. She performed alongside Queen Anne in Ben Jonson's

Masque of Blackness and later, his *Masque of Beauty*, blacked up in both, presenting 'a very loathsome sight'. Jonson went on to dedicate his play, *The Alchemist*, to her, and George Chapman his complete translation of Homer.

Mary's husband died in 1614 when she was twenty-seven, only a month after she had given birth to their first son, James. By then the pair seemed to have ironed out their initial differences, to the extent that he left her an income of £1,200 a year, plus 'all her books and furniture of her studdye and closett'. Less welcome were debts of £23,000 (almost five million at current values). The estate itself was inherited by her son. You'd be forgiven for thinking this was the end of the crisis, when in fact it was only just getting going. In 1616 her baby boy died and the estate passed from him to Sir Robert's younger brother John. While Mary had leave to remain at Loughton Hall for the rest of her life, she had few assets to generate any income and the debt remained her responsibility. Not that she allowed this to act as much of a drag upon her. She did battle with her creditors over a number of years and once, when someone applied to her father for help, he made it clear that she would allow no one else to get involved in her financial affairs. She also never remarried—a rarity at the time.

The common explanation given for Mary's refusal of marriage is that she was in love with William Herbert, Earl of Pembroke. Herbert was Mary's cousin, the son of her aunt, the Countess of Pembroke. His father was supposed to be Henry Herbert, but over the years alternative hypotheses have suggested both his uncle, Sir Philip Sidney, and William Shakespeare. By the 1610s Herbert already had a reputation as a formidable philanderer, having been first locked in Fleet Prison and then banished from court by Queen Elizabeth for impregnating one of her younger ladies-in-waiting and refusing to wed her. Undeterred, he had gone on to marry for money and influence and continued in much the same way as before. He was one of Queen Anne's favourites, a renowned literary and theatrical patron and definitely a man who couldn't keep his penis in his pants. He has that look about him in his portraits—hangdog, wet-eyed, the kind of rake who takes women to bed by persuading them they need to save or protect him.

No one knows when the relationship between Herbert and Mary began, but they had stayed in the same houses together from a very young age and there has been a suggestion that Mary's firstborn, James, may actually have been fathered by William rather than by her husband, who died of gangrene *in pudendis* very shortly after the boy was born. What is certain is that sometime in the five years after her husband's death, Mary gave birth to twins, William and Catherine, and they both took the Herbert surname. William the younger was to die fighting in Ireland. Catherine married a vicar and there's a wild claim that she may have been the mother of the Restoration novelist Aphra Behn—another feminist icon and mystery woman—though no evidence at all has been uncovered to support this (add to it the earlier claim that Catherine's father, William Herbert, was fathered by Shakespeare and you end up with the Bard as Behn's great-grandad. There is room for a forest of genealogical supposition in the absence of fact).

What's the importance of this? Mary and William's relationship closely shadows that of the central characters of *Urania*, the cousins Pamphilia and Amphilanthus. It's hard not to read the book as a kind of roman-à-clef with a missing key, and many commentators have tried to provide it. But Wroth quite deliberately breaks up her own life and scatters it across the book, appearing as Belizia, married to a forest lord and living a life of 'cold despaire'; as Bellamira, a widow whose only son dies and who is betrayed by the man she loves; and as Lindamira, faithful to the queen but betrayed by gossip, 'her honour not touched, but cast downe, and laid open to all mens toungs and eares, to be used as they pleas'd' (a brilliant evocation of the lasciviousness involved in the male shaming of women which for some reason makes me think of the *Daily Mail*). And that's before you get into the representations of the likes of Baron Denny, or Frances Howard, popularly viewed as a witch and murderess at the time but represented by Wroth as a victim of men's treachery.

After the whole *Urania* debacle, Wroth's time as a fashionable member of the royal court was over. She saw out the remaining thirty years of her life in Loughton, unmarried, petitioning the king over and over again to be protected against her debtors, finally dying with the Civil War rumbling around her. The story in its most common form is that Herbert didn't want her, and if she couldn't have Herbert she wanted no one else, that any respectability bestowed upon her by her maiden name had been squandered first by her affair and the children that resulted from it, then by the production of a scandalous, salacious text: 'she takes great libertie or rather licence to traduce whom she please, and thincks she daunces in a net'. There is, however, another reason for the ostracisation of Wroth, which brings together the personal and the political in the magical or delusional compulsion of the writer. But to understand this fully, first we have to understand Mary Wroth as Mary Sidney.

•

The Sidneys, it needs to be appreciated, were Protestant activists. Since the time of Elizabeth, they had been closely allied to Robert Dudley, Earl of Leicester, firmly at the vanguard of Reformation tendencies in England. Sir Philip Sidney had, after all, died fighting the Catholic Spanish in the Netherlands (for many years he appeared in history books as the exemplar of the English gentleman, having refused medical help in favour of others, a story which ignored the fact that he wasn't all that seriously injured until he got gangrene). Sir Robert Sidney, his younger brother and Mary's father, was later appointed governor at Flushing and annoyed the queen by organising and leading unauthorised forays against the same enemy. Although Robert Dudley was gone, the Sidneys— and the Herberts—continued their agitation into the reign of James I. When James's daughter, Elizabeth Stuart, married Frederick V, the Elector Palatine, in 1613, Sir Robert—by now Viscount Lisle—was among the party who accompanied the couple back to their new palace at Heidelberg. The Palatinate was a principality in Germany which had been, since the time of Frederick's great-grandfather, staunchly Calvinist in outlook. Indeed, Frederick was leader of the Protestant Union, a kind of mutual support group for the German Protestant states. The militant wing of Protestantism took the marriage as a sign that James was preparing to side with them. This, it turned out, was a mistake.

When Frederick was elected to take the crown of Bohemia by Protestant rebels who no longer wanted the post to be held by a Catholic Hapsburg, the theory was that he came to the throne with the support of England and the Protestant Union. He had neither. He is known as the Winter King, because this is how long he supposedly reigned. In fact the poor sap managed four whole seasons. He was crowned on 4 November 1619 and the Hapsburg army finally routed his troops at the Battle of White Mountain on 8 November 1620. He fled to The Hague with his wife and family, to live out the rest of his life in exile and shame. Electors were so-called because they had a vote in the election of the King of the Romans or Holy Roman Emperor. While this post was traditionally filled by a Hapsburg and hence a great source of power for hardcore Catholicism, the Protestants hoped that Frederick's Bohemian adventure would tip the balance, giving him an extra vote and persuading others to come over to the Protestant side. It achieved nothing of the sort, instead acting as the trigger for the Thirty Years War.

William Herbert was particularly disgusted by James's inaction, telling Frederick's representative that he considered it a failure of duty on the king's part. In *Urania*, published only a few months after these events, his avatar, Amphilanthus, is voted King of the Romans. It's hard to understand what Wroth thought she was doing with this particular move, or to imagine how controversial it would have seemed at the time, how provocative. Herbert certainly wouldn't have thanked her, not only for implicating him in the Bohemian crisis but also suggesting some kind of will to power on his own part. The mania of words—cheap spells, wish fulfilment, mischief, a kind of violence, even—all of it seems to have run away with her. The relationship between the two bit the dust shortly thereafter, lovers lost forever in the woods.

I am not a visitor to stately homes except under the most extreme of sufferances. There is something very boring but also quite upsetting about witnessing the appalling taste of the upper classes. But there are three paintings of Mary at Penshurst, the Sidneys' country seat in Kent, and I thought I should go and see them, even though I wasn't entirely sure why. You might have thought it would make more sense to go to Loughton Hall, where Mary lived for most of her adult life, but there was nothing left of her there. Instead I had to drive all the way down here and commit myself to some pantomime of *research* when I would much rather have been messing about in the trees.

It was everything I'd hoped for, and less. The school holidays were still a way off when I visited, which may explain why there were so few people present under the age of sixty. Men in unfetching but practical headwear with Reactolite glasses and expensive photographic gear lumped round the grounds with their wives—bright and stylish-looking, definitively better off once their husbands were dead. The gardens— a series of rooms open to the sky—were perfectly kept, all pinks and purples, precision-cut bushes and dodgy statuary, the Sidney emblem (apparently a broad-headed arrow called a pheon, though looking more like a thistle) embossed on anything big enough to make it look impressive. The staff seemed to have been recruited direct from finishing school then carefully aged in a cellar for fifty years, like fine wine. The house was a huge, buttery sandstone faux castle, rambling through styles and historical periods as it moved away in each direction. The whole site was designed to make you react in a certain way, to show you your place. I found myself well-spoken and polite to the point of caricature, a quivering wretch, a middle-class serf. What I didn't like about stately homes was me. The room with the pictures in was very dark and I had to keep moving around and squinting, trying to find somewhere to stand where some light showed the faces without reflecting off them in a dull, bulb-shaped glare. The nice attendant lady engaged me in Wroth gossip, all of which I was unequal to. I smiled and nodded and said sorry a lot.

The first painting shows Mal aged about nine, with her mother, Barbara Gamage, and her five younger siblings. Gamage is grouped with her four youngest children, one hand on the littlest and the other on the shoulder of William, her heir. The two older girls stand slightly separately, holding hands. The younger of the two seems uncomfortable but Mary looks straight out of the picture with a small smile of pride or amusement. Gamage herself is hunched up inside her dress, serious, ducking beneath the 'roof' of the picture. The two youngest standing children look like dwarves. William is a girl with a sword. Although the point of the sword comes across the front of her dress, something about the perspective of the picture is wrong, so that Mary, even though she is neither, seems to be at the front and centre.

The second picture is a painting of Mary with an archlute or, more accurately, a painting of an archlute with Mary. The instrument is actually taller than her, looks slightly rickety and as if it would be nigh on impossible to tune. She is dressed as Noddy Holder. With the concentration expended on the instrument and Mary's clothing, the portrait is more about what it signifies than an individual. Perhaps this is true of all art of the period. Or all aristocratic art. Or all commissioned art. Mary's face looks as if it's been painted in later. She seems very young, furtive, both embarrassed and gratified by the attention. The discomfort of her pose is eloquently adolescent.

It is indeed possible that her face was 'painted in'. The first portrait was by Marcus Gheeraerts the Younger, this one by his brother-in-law, John de Critz. Like the aristocracy they painted, the Flemish painters who worked the court circuit, having fled persecution for their Protestantism in the Netherlands, tended to marry into each other's families. They also employed assistants and left them to do the grunt work of dresses and finery, leaving their own talents free to concentrate on the faces. This is what Marcus Gheeraerts was renowned for—bringing a new level of realism to the features of the great and good of late Elizabethan and early Jacobean England. It is unmissable in the last of the three portraits, of an adult Mary and her mother. It's a picture of sharp contrasts of skill and style. The dresses are bizarre, the constriction of Mary's waist so exaggerated that it looks as if she's being sucked into some sort of whirlpool or vortex (her skirts). Her cleavage is a slapped-on black line, like a cartoon bum crack. But the faces are presented in a naturalistic style, pale, greyish. Mary's mother stands behind her at her shoulder looking pensive or distracted. Mary looks straight at the artist, apparently half-smiling, even though her mouth and eyes are almost identical to her mother's. Only her nose distinguishes her, ever so slightly large with a definite change of direction at the bridge. It's the nose, still somehow presented as delicate—that's Gheeraerts's genius right there—which animates the smile and gives her character. Or that's what I decided. As a possessor of a large nose I am a great believer in how they add character. And what is looking at portraits of people you want in some way to *know* but an exercise in projection, anyway? What was any of this but an exercise in projection?

•

I couldn't shake the idea that the text of *Urania* was some-how *encoded*. It's what had made me think of Roussel, who in effect encoded language within language. The traditional view of Wroth—partially born out of the feminist perspective through which her work has been rediscovered—is that she encoded the frustrations and disappointments of her own life, and hence of women. But this doesn't quite hold: she made very little effort to disguise the narratives she was drawing from the social world around her, hence Baron Denny's anger. There is something else going on in the text, something a little deeper, or a little stranger, anyway.

The world view of the Sidney family was a mixture of Protestant Christianity and the various traditions of Renais-sance mysticism which, as Frances Yates has shown across a number of hugely influential books, congealed together in the late sixteenth and early seventeenth centuries, a mixture of Hermetic magic, Christian Cabala and Paracelsian alchemy. The Hermetic tradition is traced back to the works of Hermes Trismegistus, held to be a priest in Ancient Egypt though later proved to be nothing of the sort. Christian Cabala is basically a wholesale appropriation of the Jewish Cabalistic tradition but used to Christian ends on the basis that if you monkey around with the letters enough, the name of Jesus appears embedded in the original text of the Old Testament. Alchemy had a long and often disreputable past but viewed through the works of Paracelsus became both a source of new medical knowledge and a higher search for the transformation of self, a kind of spiritual quest. What drew these ideas together was a Platonic conception of the universe as a perfect reflection of God at every level, from the Cosmos to the inner workings of the human mind. Sir Philip had been taught by the key occultist of the Elizabethan era, Dr John Dee, and Mary's uncle's library of books was held after his death at the Sidneys' home, Penshurst in Kent. Mary's aunt, the Countess of Pembroke, was herself described as 'a great Chymist'. It's even arguable that Mary's nephew, Algernon Sidney, drew on the same Platonic conception of the universe a whole generation later in order to argue for republicanism: 'As death is the greatest evil that can befall a person, monarchy is the worst evil that can befall a nation.'

With the Protestant excitement around Frederick V and his marriage to Elizabeth Stuart, these same mystical influences welled up on the continent and resulted in some of the most famous texts of the time, the Rosicrucian manifestos. Taking the inspiration for their mythical leader, Christian Rosenkreutz, from Spenser's Redcrosse Knight, drawing heavily on Dee—who had travelled extensively in Germany and Bohemia—these texts can be viewed as a last flowering of Renaissance Puritan mysticism. Across the *Fama* and then the *Confessio*, the Rosicrucian authors presented themselves as a secret society or brotherhood established by Rosenkreutz and building on his studies in Europe, the East and Africa, 'so that finally man might thereby understand his own nobleness and worth, and why he is called Microcosmus, and how far his knowledge extendeth into Nature'.

The works were known in England, as is evidenced by Robert Fludd's *Apologia* for the Rosicrucians in 1616. It seems likely that Mary Wroth would have come across them, especially considering her role as one of the chief conduits—through her letters to and from the king's ambassador in the Netherlands, Sir Dudley Carleton—for information about the developing crisis in Bohemia. But the defeat of the Winter King at the end of 1620 is a turning point for the Renaissance conception of the universe. James I was deeply suspicious of magic in any form, as the book he wrote in 1597, *Daemonologie*, makes clear. Indeed Fludd was brought before the king in 1617 to defend his defence of the Rosicrucians from claims of sedition, something he managed with such chutzpah that he ended up as the king's doctor. A witch scare and a panic about the Rosicrucians swept mainland Europe in 1623, the same year that Marin Mersenne launched a blistering attack on the ungodliness of the Renaissance magic tradition, making in the process, Yates argues, the necessary intellectual space 'for the rise of Cartesian philosophy' which ushered in the Enlightenment. As well as being a monk, Mersenne was René Descartes's agent. Rubbishing a whole world view so your client can make a splash—that's earning your 15 per cent.

The third and, to me, most interesting Rosicrucian text is *The Chymical Wedding* by Johann Valentin Andreae, from 1616, probably because I prefer stories to manifestos. In it Christian Rosenkreutz recalls his invitation to a strange and mystical royal wedding, although we never learn who the married couple are. The book is taken up with a decidedly Roussellian description of the seven days of revels, all portals and strange feasts, caves, weighing sessions, plays and coffins. Of course the clue is in the title, and the book is 'an alchemical fantasia, using the fundamental image of elemental fusion'. Like alchemical texts before it, it explores—or sets out in metaphorical language—the method of creating the Philosopher's Stone (also known as Our Mercury, philosophical Mercury, *mercurius duplex* and by many other names). To put it another way, it's not supposed to be read literally but symbolically.

Once you start to look at *Urania* in similar terms, the book comes into a different—though admittedly a squinting—kind of focus. It certainly makes much more sense as an alchemical recipe of some sort than as anything we would recognise as entertainment. Either Mary had missed Ken Campbell's trick of leavening things with a joke, or the humour was a little too rarefied for me. Alchemical procedures, however, usually end—the Philosopher's Stone is made. If you think of each character in Wroth's *Urania* as a different substance (or as a different aspect of a few substances, represented by the many different kingdoms) then Wroth seems embarked upon an endless shuffling, a forming and breaking of compounds that never finishes. This is less a recipe than a never-ending series of experiments. There's a sense that she has only inserted the elements of a roman-à-clef, the shadowing of her own life and those of the people around her, to distract the everyday reader from this deeper purpose. However, as everyday readers—and ones way out of time—it's hard to be sure what that deeper purpose was. Magic, the steady accretion of ritual and hidden meaning, the systematic application of drifting metaphor, is difficult, it is abstruse and, yes, hermetic. We may dig our fingernails into the bark of the text, we may even make a lunge for the branch just above our head, but we don't have the tools, the skills, the experience of *this particular kind of tree*, to get ourselves off the ground.

If you head away from the Doom Tree, twenty metres or so downhill in the direction the path runs, you come to another huge beech. Anywhere else in the forest it would be the one that catches your eye, but next to the Doom Tree it looks almost ordinary, graceful and self-effacing. Although it probably isn't any less cut into than the Doom Tree, it seems to deal with it better, with less anger and histrionics, a slight aloofness. Its strength feels contained, channelled to the sky. To pull yourself up into it, you have to pass beneath a strange growth which presents as a cross between a lamb embryo and a sexual appendage. It makes the short, breathless scramble feel like crossing a threshold into another world. The bark up there is very smooth, almost shining on a sunny day, the texture of raw silk. A long, wide beam moves gradually up and away in front of you. Carefully carved in, just beyond your trembling hands, is the symbol of the Illuminati—a pyramid with a human eye inside it. Below it, though upside down, someone has scraped out rectangular letters saying 'LOVELY'. It looks like it should be a restful place to set yourself down, but there's something about the topography which makes you feel unbalanced however you try to position your weight. Contrary to my enantiodromic theory of tree-climbing, every side of the Lovely Illuminati tree, you quickly learn, is the best side for falling out.

CHAPTER 4

North Weald | Ongar

I t was the sign that stopped me. I was cycling up a hill along a gravelly, potholed track, sweating slightly, breathing in a manner which made me sound more excited than I actually was, trundling past some sort of military-industrial building or demolition site, having a nice enough time until, as I began to freewheel down the other side, I came to it. In red letters on a flaking white background across a piece of corrugated iron, it read:

PRIVATE ROAD

NO PUBLIC FOOTPATH

NO BRIDLE WAY

It should already be clear to you that I'm not naturally a crosser of boundaries. Or rather, director-I thinks of himself as such, but actor-me struggles to make good on that fantasy. I'm scared of dogs and of being shouted at. I'm frightened of causing offence and of being unpopular. People I don't know make me nervous. Similarly, the countryside in general—all those animals and angry men with guns and grudges. And I cannot reiterate this enough. I am scared of dogs. The barking in particular, also the jumping up, the chasing of things, inanimate and animate, and, yes, the biting, both real and imagined. Dogs have too many teeth. Dogs are drooling, barely tamed wild beasts. They always say, *He's just playing, he won't do you any harm.* I am frightened of dog owners, and a little angry with them too.

I braked and my bike stopped. The sign was surrounded by blackberry bushes so I ate a few, sweet and cold, and looked down the track towards a big, ugly house squatting among some trees. I appeared to be at the highest point of a wide plain. In the distance, wheat was being harvested, huge boxy machines working up and down fields, hazy in clouds of yellow-grey dust. Buggies whirred around like clockwork toys on what must have been a golf course. The sky was a pale, duck-egg blue cross-hatched with contrails. I could hear the roads in every direction but there was no sound near to me. The field beyond the track was punctuated by a series of concrete groynes washed up miles from the sea, part of some latter-day megalithic temple. I looked at the sign again and tried to force myself past it. It was hand-painted by someone with rudimentary brush skills who nevertheless eloquently expressed the oceanic rage boiling inside him. I ate some more blackberries. I turned and pedalled back up the road.

After my visit to the psychic I'd tried my own feeble attempt at magic when I typed out and then chopped up the words she said, drew phrases randomly out of a hat and stuck them to a piece of paper. At the top it said: 'ONGAR

RAILWAY / TRAGEDY AROUND THE LINE'. I already knew the magic hadn't worked, but in the absence of anything better to do I decided to treat it as a spell. It was the only part of that map I followed before things started moving in their own direction, which is another way of saying it worked. I took a tube train out to Epping with my bike, thinking I would follow a course from the station along whatever road kept me closest to the railway line. I didn't have a proper Ordnance Survey map with me, just a piece of paper—which was my first mistake—and the road seemed determined to sheer off and away from the tracks, keeping me at a distance from any tragedy, real or imagined. I was heading out of the woods, into the nominal or spirit forest, where the trees had been pulled out and turned into fields well over a hundred years ago. If my understanding of the thirteenth-century boundaries of Waltham Forest was correct (the chimney containing the smoky tendril of the current Epping Forest), the route I was taking, east from the town of Epping, was a wiggly, approximate perambulation of the northern border.

I wasn't supposed to be following the Ongar railway line for the sake of it, of course. I was supposed to be looking for *tragedy* along the line. We use that word now, via news bulletins from war zones and train crash sites, to mean a major disaster, any situation in which a large number of people have died, usually in some needless or futile manner. Its roots in drama are a little more complex, a little more charged. In *Poetics*, Aristotle claims that a tragedy's purpose is to bring about a kind of catharsis in its audience, by which he seems to have meant some quasi-medical purifying of negative emotions through their expression: 'by means of fear and pity bringing about the purgation of such emotions'. In the late nineteenth century, Friedrich Nietzsche devoted a considerable amount of his time to arguing with the dead Ancients and, unsurprisingly, he saw it rather differently. For him, the point of tragedy wasn't to *purge* yourself of pity and terror at all, but 'to celebrate oneself the eternal joy of becoming, beyond all terror and pity'. Nietzsche felt that the later Greeks had lost their grip on tragedy because they became too much in thrall to the sun god, Apollo, and turned their backs on Dionysus, the god of fruitfulness and vegetation, of wine and ecstasy. For Nietzsche, re-embracing the legacy of this forest deity was a matter of 'saying Yes to life even in its strangest and most painful episodes'. The strange, the painful, the pitiful, the terrifying are all a part of life, I take this to mean. To purge them from life is to empty life.

In his book *The Chemical Theatre*, Charles Nicholl draws on the work of Frances Yates to apply an alchemical reading to *King Lear*. 'Tragedy,' he says, 'is … a drama of redemption. Its unremitting depiction of loss and doom is placed in a wider context of renewal.' He then sketches out an unlikely connection, claiming that tragedy follows the same formal pattern as alchemy: 'Both alchemy and tragedy define a journey: a road to wholeness that goes by way of dismemberment and dissolution.' Dismemberment interested me. I was riding the edge looking for a ghost, a dead knight with his head tucked under his arm, a man called Wally Hope.

I t was due to the No Entry sign that I ended up on the bridleway, cycling through a field. It was hot and I didn't want to go all the way back to the main road. Anyway, I was pretty sure the Ongar line lay just ahead. Not that I was dealing with some obscure antiquity here, some legendary thoroughfare waiting to be unearthed. It was part of the Central line up until 1994, with tube trains running up and down it. Not many tube trains, admittedly. Single-track with one passing place, it was never busy. In fact, it was claimed at the time that each person using the service cost London Underground eight pounds for every journey made. Local rumour suggested it was only kept open in order to carry the Cabinet out of the capital to their bunker at Kelvedon Hatch in the event of nuclear Armageddon. By 1994 the threat of such a cataclysm had obviously passed, or some clever cost–benefit analysis had proved that we'd be better off with them all fried in Westminster.

I could still have continued my tragical mystery tour by train. The Ongar line is now one of those heritage/steam/Thomas the Tank Engine affairs run most weekends and some weekdays during the summer by volunteers in tight-fitting anoraks. It costs thirteen pounds, however, and I'd spent enough time surrounded by elderly trainspotters, foreign tourists and other desperate, tired families with little children when my own kids were small. In my experience, the trains are always cramped and hot, the seats smell and you are expected to smile manically at each other and wave incontinently at passers-by. Instead I'd opted for my bike. Even before I got this far I was nearly knocked into the ditch by one of the drivers of the vintage London buses which collect passengers from Epping and take them to North Weald, where the engines leave from. And I got that old-fashioned inner city experience for free.

It did play its part in my unwillingness to return to the main road, though, hence the bridleway at the edge of the field, juddering over the dried-out ruts made by horses' hoofs and mountain bike tyres on colder, wetter days. The path plunged in among some Scots pines and I found myself on a bridge crossing the tracks, which ran away in a long, straight line, narrow and with plenty of cover on each side if some-one were to launch an ambush. I wondered if there were some local military re-enactment society who could be tempted into an attack on the train people, a kind of violent, traumatic, Big Society ritual. I imagined the bus driver tied to the tracks, begging my forgiveness.

Wally Hope wasn't actually a knight, not really, despite recruiting volunteers for 'Operation King Arthur'. He was a hippie, though a strange one, at least by my own reductive definitions. He kept his hair and beard cut short and neat and he favoured military garb over beads and flowers, in particular a Cypriot army uniform and the tartan of the Black Watch regiment. He certainly saw himself as some kind of warrior. He missed the key years of the 1960s locked up in Wormwood Scrubs after his own mother busted him for drugs, so he never got to be *groovy*. His connections in London were to a group known as the Dwarfs, a UK equivalent to the Dutch Kabouters, who had themselves cut their teeth with the anarchist Provo movement. In one account of his short life, it's claimed that if he was alive now Wally Hope would have been a Thatcherite. While there's definitely something more Richard Branson than Raoul Vaneigem about a man obsessed with British Airways stewardesses (he wrote to the airline to invite their female cabin crews to the first Stonehenge Festival), that comparison sells him short. He never made a penny out of any of his ventures, not because he failed to but because he never wanted to. He wasn't packaging the hippie experience for profit. Nothing as neat as that. He was half Apollo ('EVERY

DAY IS SUN DAY') and half Dionysus ('FREE STONED HENGE ROCKS OFF'). In his obituary in the renowned British countercultural newspaper of the late 1960s and early 1970s, *International Times*, Heathcote Williams—who had worked with Ken Campbell the year before—described Wally as an 'astral commando, dressed in the uniform of an armed lover', which summons up his dichotomous nature.

I've always had a problem with hippies. At least I've always thought I do. I was born in 1969, my dad was a long-haired social worker and my sister and I were often dressed in matching Clothkits outfits, so I can't honestly claim that the whole thing passed me by, but then the *hippie* is always someone else, not you, not your friends or family. By 1977 my sister was a punk—despite not really being old enough—and the whole world, even eight-year-old boys doing the slow bicycle race at a Silver Jubilee street party, seemed to have picked up on the fact that hippies were rubbish. This general feeling, the sense of a style and outlook that was past its time, was formalised and given shape by *The Young Ones*, the TV show which ran from 1982 until 1984 and established beyond doubt, through the character of Neil, that hippies were depressed and depressing, pulse-filled farters, boring, pathetic. Looking back at it, despite its claims to be in the vanguard of a new wave of alternative comedy, *The Young Ones* was a rather conservative show, teaching us that punks were violent psychopaths (Vyvyan) and that anarchist-poets were deluded mummy's boys (Rick). For some reason it was Neil I really objected to, though. It was his passivity which riled me, probably because I recognised it as part of myself. I'm never more alive than when feeling angry about a decision I've refused to take. But then who would want to see that in themselves, to look forward to year

91

after year of this particular game? It was in the wind, it was everywhere, it was the most mainstream thought you could have: *never trust a hippie*.

The irony of all this should already be apparent but let me lay it bare. Here is a man traipsing around, climbing trees and communing with chlorophyll, all the while telling you he has no time for hippies. The writer Sara Maitland strolled through these same woods in the company of Robert Macfarlane, the dashing cavalry commander of the new nature writing movement. Delivering yet another *coup de main* (or, more accurately, a *coup de pied*), Macfarlane had suggested that the pair remove their shoes, so that they could directly experience the forest beneath their feet. Not to be outdone, feeling a weird macho deficit of feeling, one summer day I ascended the southern slopes of Loughton Camp with my boots in my hands and my socks in my boots. It was nice as long as I walked carefully— the air between my toes, dorsal hair riffled by the breeze, the unclenching of all those nerve endings as they found, for once, they'd been given something to apprehend. It was nice until I bumped into two female hikers coming down the hill. They looked at my feet, luminescent in the shadows, almost obscenely naked. I looked at my feet. We all looked at my feet. As soon as they'd gone I put my shoes back on. Was it this nudity, this openness I recoiled from?

Beyond the railway line, the track I was following led to a small wood, mainly more Scots pine, tall trunks drawing your eyes up towards the sky. At Chingford, back down in the centre of Epping Forest, the only Scots pine is a disguised mobile phone mast, the pattern of the bark repeating, like history, first as tragedy then as farce. There was a pond just beyond the gates, self-described on various signs as a private lake—'KEEP OUT'. As soon as the path went under the trees it turned to mud and quaggy puddles, churned and re-churned by walking boots, wellies, bikes, ponies and whatever fishermen use to drag their gear to private lakes: tackle trolleys or rodmobiles or angler wagons, I suppose. I sludged through it on my bike in a low gear, the sound of clumps of clay rubbing on my mudguards making a panting noise just behind me so that I felt pursued. At the far end of the wood the path split and I stopped, trying to work out which side of the hedge I needed to be. Although both paths ran along, parallel, moving in exactly the same direction, it seemed like a decision I needed to get right. I opted for the pea field and, as I moved, a buzzard took off from its perch in the canopy of the pine twenty feet above me. It called out once, twice, three times, and so melancholy and lost was the sound that I decided it was crying for its mother. It called again, a falling note, as if dropping noise back to earth. Moving with a lazy disjointedness which its voice may have encouraged me to project onto it, again it called, an old child singing blues, bending the same note down over and over, like Miles Davis doing it wrong until it's right. I didn't move again until I was sure it was gone. And nor did all the other silly birds.

I came out of the woods next to the water tower at Toot Hill, a landmark for the legions of club cyclists who pour out of London into Essex every weekend. Cruising down a narrow lane leading to the main road, I must have flickered for a moment in the peripheral vision of a horse doing some sort of dressage routine with a girl in a hot-pink vest. The horse was startled and began to bolt, the girl rocking back in the saddle. I didn't wait to check how much harm I'd done. Later

I saw a kite made to look like a hovering raptor fluttering in the wind to protect what appeared to be a lawn bowls club in the middle of nowhere. All the signifiers of wealth were on display round here, but none of them seemed to be in quite the right order. Maybe, like the buzzard and Miles Davis, the local residents keep on doing it wrong till it's right. Either that or I just didn't get it, couldn't read it properly, though I was loath to admit that. I grew up in a village on the edge of a city, albeit no London. I spent my summer holidays trudging through fields like these—perhaps flatter, perhaps not so picturesque—hoping to find something to do, waiting to be chased off by farmers. As in these Essex–London borders, our local MP was a Thatcherite headbanger. By rights I should've felt at home. Though to be fair, I'm not sure I ever felt at home at home, either.

Wally Hope, aka Phil Russell, was the founder of the Stonehenge Free Festival, the longest-running free festival in the UK, the least commodified, most unpredictable bacchanal of late twentieth-century Britain. The first event was in 1974 and it carried on annually until the police, in a quite shameful attack, broke it up in the so-called Battle of the Beanfield in 1984. Along with Orgreave a year earlier and Hillsborough five years after, this incident—with allegations of doctored police radio logs and photo negatives going missing from national newspaper archives—will eventually take its place in the 1980s pantheon of police excess and cover-up. The first festival in 1974 was a long way, though, from this eventual crisis. On this inaugural occasion, just a few hundred hippies turned up and were entertained by one live band, the synth act Zorch.

Nobody knows where the use of the name *Wally* came from. There's talk of a lost dog or an AWOL sound engineer at the Isle of Wight festival. Sometimes Wally Hope would say it was Gaelic for *elite warrior*, which doesn't appear to be true. At others it was supposed to be an acronym for the Wessex Anarchist Libertarian League of Youth. Whatever its root, Wally adopted it—either as a tactic or out of a genuine desire to change the world—as one of his articles of faith: 'I look to the revolution to rename every citizen with one sound and the composite name of all citizens to be the analogue of the deepest terrestrial vibrations so that when we are all called we will all hear.' The sound was *Wally*. By the time I came across it, growing up in Leicestershire, it had come to mean any hapless, useless idiot, but there we are.

Free festivals had been done before. The Isle of Wight, like Woodstock, had been declared 'free' after people pulled the fences down. The first Glastonbury Fair in 1971 was a free-entry event, though highly organised by the standards of the day and funded by wealthy benefactors. The Windsor Free Festival in Windsor Great Park started in 1972 on the Queen's land because, in the words of one of its key organisers, Ubi Dwyer, 'it's an effort to reclaim land enclosed for hunting by royalty centuries before'. It had been broken up by force in 1973 and again in 1974. Wally was present each time, dancing in front of the police or charging their lines—depending on the account—until they clattered him. What distinguished the Stonehenge Festival, though, was that Wally had no intention of leaving.

In the summer of 1974, when the very first Stonehenge Festival finished, Wally and about thirty other Wallies stayed on. This had been Wally's plan all along. There was Phil Wally,

Chris Wally, Arthur Wally and Wally Egypt; there was Sir Walter Wally, Wally Moon, Ron Wally and Scots Wally. There was even Wally Woof the Dog. The Wallies gave their encampment an address: *FOGHT WALLY, c/o God, Jesus and Buddah, GARDEN OF ALLAH, Stone Henge Monument, Salisbury Plain, Wiltshire.* The previous summer, Wally Hope had declared his mission in the *Windsor Freep*, the bulletin handed round at that festival: 'Our job and meaning in life is to create heaven on earth.'

Salisbury Plain turned out to be a hard place to create heaven. As summer turned to autumn, the Department of the Environment moved to evict the Wallies. The press descended to get a feel for the life of these 'sun worshippers' and filed stories about apathy, dope-smoking, cold, relentless winds, driving rain and inedible stew. Wally wasn't always there to poeticise the experience, to render the banal mythological. Often he was up in London preparing the Wallies' legal defence, which revolved around the idea that the Chubb family (yes, of the locks) had left the monument to 'the Nation', not to the Department of the Environment and hence that it was a kind of common land. Oh, and also that God had told the Wallies to go and live there. Perhaps unsurprisingly, this appeal to divinity didn't win the day in court, but the case offered Wally a platform and he took it. After the verdict went against the Wallies he left court to claim 'we have won … we were playing with the ace of hearts. The judge told us we were a hundred per cent good people. We are friends with the judge. We are even friends with the other lawyers. We have made friends with all sorts of people. So we have won.'

When the bailiffs were sent to move them on, the Wallies shifted their encampment a few feet onto neighbouring land and it was winter rather than hired security that closed down

Foght Wally altogether. Then Wally began handing out leaflets for the following year's Stonehenge event, driving round London in a rainbow-coloured car with his teepee strapped on top, wearing a kilt and a top embroidered with the eye of Horus, once again telling anyone who would listen that he intended to stay there, to live there, to create a permanent bubble of freedom, a heaven out on the Plain.

I stopped at the church at Greensted. It's a small, wooden thing, ridiculously quaint, with notices telling you what to do about everything and a twelfth-century Crusader buried out front. It was very dark and very warm inside. Two women were doing some admin in the back room—moving hymn books or stacking vestments or whatever you do in churches. One was plummy, the other less so. The light flooded out of that back room yellow and dusty and smelling of damp books, and I wanted to walk into it and have a cup of tea and sit quietly in there and be saved. The plummy one had just been to see the rugby sevens at the Commonwealth Games and came across the South African team back at the airport. She was surprised at how small they were in the flesh. The other less plummy one did what less plummy ones do and murmured her assent. The walls were plain, dark wood, the lower panels crushing up together like the teeth at the front of my uncorrected jaw.

A small old man with a hearing aid was sitting in one of the tiny pews, trying to catch my eye. I attempted to avoid being caught because churches are another of the things which make me feel uncomfortable and I both don't believe in any of it and don't want to cause offence. The room was very small, though, a kind of toy church where a miniature South African rugby team would feel right at home, their perfectly formed, doll-like knees butting up against the cool wood of the bench in front of them. My eye was trapped. It had nowhere left to go. First he told me that this was the oldest wooden church in the world, which I kind of already knew. Then, sensing that this hadn't hit the spot, he pointed to a fist-sized lump of sunlight coming through the wall and informed me it was a leper hole, where sufferers could have their Communion wafers passed out to them. Sensing that gore combined with mystery would finally sate me, he pointed up to a relief carved on one of the wooden roof beams.

It showed a man's crowned head with raggedy neck. An animal, perhaps a chihuahua of some sort, was crouched down near the neck, sniffing or licking at it. When I asked the old man what it represented he said he didn't know, but the plummy woman breezed in to tell me it showed St Edmund being protected by a wolf in the forest somewhere in East Anglia. Although the poor man died in 869 his body rested here on its way to or from Bury St Edmunds in 1013 (what does *body* even mean after 144 years?). The backstory involves one Ivar the Boneless, a Viking with a comic-book name and, presumably, extremely bendy limbs. Ivar and his Great Heathen Army (really, you couldn't make these up) defeated King Edmund—who made the move from coronation to beatification only after bifurcation—and had him beheaded for refusing to renounce Christ. The king's head was then thrown into the undergrowth for Great Heathen tree spirits to play football with. When Edmund's followers went to look for him and shouted his name, a wolf called back. And there it was, a huge grey wild dog sitting guarding the king's unchewed bonce. The carving didn't necessarily convey the majesty of this tale. I've already mentioned that the wolf looked more like a toy dog, terrifying in its own way but not exactly lupine. As for Edmund, his eyes were wide with surprise and horror and his mouth was round as a sex doll's. It wasn't the right tragedy, anyway. Already I felt myself a connoisseur.

In May 1975, Hope was spending the night at a Wally squat in Amesbury when the police raided the house, supposedly looking for an army deserter (which, incidentally, meant they didn't need a search warrant). Instead, they found three tabs of acid in Hope's wallet and promptly arrested him. He was refused bail and held at Winchester Prison. When he declined to wear the prison clothes, which were giving him rashes—despite the fact that remand prisoners are supposed to wear their own clothes—he was dragged off to see not a dermatologist but a psychiatrist. He was promptly diagnosed with schizophrenia and allegedly given quantities of Largactil and Modecate way above the recommended doses (up to twenty-seven times the recommended daily dose of Largactil according to one account). Unsurprisingly, this chemical beheading turned him into a shuffling wreck, into exactly the kind of malfunctioning human being he was supposed to have been to begin with. He was sectioned and transferred to Old Manor Hospital in Salisbury, where his new psychiatrist took his current state as a baseline and continued with the same *treatment*, albeit at reduced—though still very high— dosages. A group of friends worked out a plan to get him out of the hospital and on a boat to Europe but when they put it to Wally he refused, preferring instead his martyrdom. This was his crisis, the point where he could have chosen life, and Wally didn't manage to.

The second Stonehenge Free Festival went ahead without Wally and this time thousands of people turned up. He was released three days after it finished, with no explanation, now apparently cured. He was so broken by the experience, by the drugs, that it took him over two days to drive the one hundred and thirty-odd miles back from the hospital to Essex, stopping every twenty minutes or so in a layby to rest and recover before carrying on. He was heading for Ongar House, the home of Ted and Sylvia Hatfield. The Hatfields were both doctors and had established their practice in Chipping Ongar straight after the war. The town sits at the top right (or north-eastern) corner of our chimney. The Hatfields were old friends of Wally's father, who had died when the boy was only twelve, and when Wally had come out of Wormwood Scrubs the terms of his parole had been that he would live with them. He had stayed there on and off ever since, sent them cheerful postcards from Cyprus which he signed off, *Your sun.* They were family.

The two GPs were shocked by Wally's condition on his return, fearful that the huge quantities of antipsychotics had caused chronic dyskinesia, an incurable condition of tics and shakes, of dull eyes and swollen tongue. Wally rallied enough to visit friends in Dorset then travelled on to the Watchfield Festival, where he gave a videotaped interview in which there was a suggestion that he was intending to buy land in order to further his plan of creating a heaven on earth (or at least, that other people were hoping to persuade him to). When he returned to Ongar, 'I was very worried to find that there was little change in his condition,' wrote Dr Hatfield. 'He was still walking about like a "Zombi" [sic], much the same as when I saw him in Winchester Jail ... He appealed to me for treatment, but there is no treatment for this condition…Three days later he was dead.'

While I was in the church at Greensted, clouds snuck up on me, and the darkness only added to the sense of incipient gloom when I reached Chipping Ongar, less than a mile further east. In contrast to Epping, which still seems like an affluent market town and has chain coffee shops and a Boots and so on, Ongar felt a bit run down. There was a poster calling for HOME RULE glued to the back of the parking sign but the main sense of the place was one of withdrawal. It was hard to imagine anyone bothering. It was half four on a Friday and almost everything was shut. The newsagent had a handwritten sign in the window saying 'NO HOODIES'. The swanky hairdresser's was run by a Gary Pellicci, and I wondered if he was related to the Pelliccis of Bethnal Green, who used to feed the Krays. Ongar feels like the kind of place unimaginative East End villains would come to hide out at before easyJet made it cheaper and quicker to go to Spain.

I walked once up the road and then back down, during which time the café where I hoped to get a cup of coffee closed. There wasn't much to do except stand outside Ongar House gawping and taking photos on my phone. It's a very grand building, fourteen windows at the front, each of twelve panes of glass. Like the rest of Ongar it looked grubby and past it, though, paint bubbling and flaking away from all those windowsills, the bushes overgrown, the bricks very grey under a grey sky. It was here, in the kitchen at the back of the house, in 1975, that Wally Hope—or Phil Russell, as the Hatfields would've known him—died, choking on his own vomit. There was to be no heaven on earth, no earth in heaven, no more Sun Days, no victory for the astral commando. His heathen head was kicked off into the trees and no animal came to protect it.

'By means of fear and pity bringing about the purgation of such emotions.' I needed some help with fear and pity myself. Or with fear and self-pity, anyway. I was developing an idea, an inkling, that the roots of my current problems all lay in decisions I had taken—or more often, ducked taking— earlier in my life and that these decisions or non-decisions had been based on fear. That I was in effect a coward, a runner not a fighter, exactly the kind of person you didn't want at your back in an emergency. I'm not going to embarrass myself with examples of my cravenness, you'll just have to take my word for it, while knowing full well that this is an example of it too. I feared big things (the end of the world, ecological collapse, war, famine, illness, death) perhaps less than I should, but relatively small things—offending a stranger, being barked at by a dog, being told off, being *disliked*—with a kind of damp-palmed horror. I thought I needed catharsis, and later I found it, though you'll have to decide for yourself what kind it was. For now I'll leave you with this: a dead hippie, twenty-eight years old, lying in the dark on the kitchen floor of a house in a market town in Essex, his throat and lungs blocked by vomit. Me, standing outside on a gloomy Friday afternoon almost forty years later taking pictures on my phone. Wally Hope was not the only dreamer, not the only man to work at a tangent to the central thrust of English culture, not the first and certainly not the last to end up like that. Nor, I'm sure, was I the first to find a cautionary element to this short, bright, muddled life, a reason to stop and turn back when the sign says 'NO ENTRY'.

CHAPTER 5

Lippits Hill Lodge | St Paul's Chapel

The car had collided with a tree, rearing forward over a slight bank and into a small ditch, its bonnet pushed back into a surprised snarl, its detached bumper a gumshield dislodged by a knockout punch. It wasn't a large tree, just a scrap of a thing, an ash with its roadside branches sawn off, but it had won this particular encounter. Stuck to a side window was a police notice to let passers-by know that they knew the car was here, an exercise in epistemology instead of law enforcement. I looked back up the long, straight, quiet country lane and tried to imagine how anyone could have managed to pull off this particular feat of ineptitude. Some small part of me found it impressive. On the other side of the road was a wall, maybe eight feet tall, with a kind of folly watchtower or turret at its end before a large, black, spiked gate. It was the kind of border Prince Charles loves, supposed to look old but built out of bricks that will never age. An engraved slab of sand-coloured concrete confirmed that it was the entrance to Lippits Hill Lodge. The connection between this and the crashed car was almost too good to be true. It was as if Wally Hope had nodded out once more on his long drive back to Ongar and abandoned his car to ask for help—of all places!—at that gate.

Wally's painful, slow, disoriented two-day car journey from Old Manor Hospital in Salisbury back to Essex in July 1975 had become twinned in my mind with John Clare's 'Journey Out of Essex' in July 1841. They covered roughly the same distance, one heading east and the other north, one returning to Essex and the other escaping it, both fleeing mental health institutions, acting as distorting mirrors for each other's plight. Clare's story is better known than Hope's, constituting as it does a large part of the myth of *poor John*, the peasant poet turned mad genius. And Clare's journey started here, in the building somewhere behind the gate. Back then, Lippits Hill Lodge was an asylum.

He had been there for four years, from 1837 until 1841, and thought his friends had forgotten him. He missed his wives, both his real one, Patty, and the childhood sweetheart, Mary, who had become his muse, the 'hopeless hope' of his incarceration. He missed sex. Most of all he missed freedom—limited, compromised, painful freedom. Meeting an old Gypsy in the woods, he decided to escape. The Gypsy offered to help but became less interested when he discovered that Clare had no money. When John returned to the camp a few days later the Gypsy was gone, leaving only a hat. Clare was not deterred. He took the hat, put it on his head and walked the eighty-odd miles back to Helpston in Northamptonshire, living on grass and later chewing tobacco to keep hunger pangs at bay, lying down at night with his head pointing towards the pole star so he would know which way was north when he awoke.

It took him four days and three nights to complete the journey and when he got there he was told that Mary, the 'first wife' of his imagination, had died during his incarceration. He went on to the cottage of his actual wife, Patty, at Northborough, whom he barely recognised. He was 'homeless at home & half gratified to feel that I can be happy anywhere' (note the echo of his 'hopeless hope' in that 'homeless... home'). Before the end of 1841 he had been dragged off to Northampton General Lunatic Asylum, where he was to stay until his death in 1864. The doctor who certified him said his mind had been weakened by his addiction 'to poetical prosing'. Of all my own vices, poetical prosing seems the least dangerous, but I may be in denial.

Eighty years later, Jacob Epstein's eldest son, Theodore Garman, would spend time in the same asylum at Northampton, now rechristened St Andrew's Hospital. It was here, just as the Second World War ended, that he was diagnosed with schizophrenia—and in contrast to Wally's case, it seems that the diagnosis was accurate. While Jacob actively encouraged his younger son Jackie to paint (and Jackie the Engine resisted), he never lived with Theo or publicly acknowledged that he was his child. Despite or because of this lack of paternal encouragement, Theo became an artist. He only exhibited three times during his life but Wyndham Lewis commended the 'vitality' of his work, 'these heavy coarse green forms, these large pale faces in which the universal green is reflected'. His paintings seem indebted to Cézanne and early Matisse and the colouring of Van Gogh. There's something naive about them, something *outsiderish*—particularly the flat, slightly vacant human faces—and their brightness is startling. They would have seemed unfashionable in the 1950s, but you could imagine him being adopted by Kitaj and others in the 1960s.

Theo, or *Sonny* as his family called him, became increasingly sick over the next few years, walking leaning to one side, muttering to himself. He took to drawing still lifes of flowers and small religious statues. There's a strange reductionism to making art of art, as if he thought he could recapture the divine through his representations of these relics. Theo had borrowed, or possibly stolen, one of the statues he was using from Chelsea College of Art and when the school authorities threatened to prosecute him in 1954 his mother, Kathleen Garman, decided he needed to be hospitalised again. He was invited to dinner at Epstein's house in Chelsea, then taken against his will, restrained and sedated. The sedation doesn't seem to have gone as planned. Only thirty years old, Sonny had a heart attack in the ambulance and was dead before he reached his destination.

There's another journey to mention, which in some way runs counter to the 'escapes' of Clare and Hope (though it continues the rough direction of the latter's and the former's mode of transport). You could view it as part of a trinity, an exploded triangle, the three routes scrawling a Ouija board 'Y' across the topography of southern England. In 1599, William Kemp travelled right along the southern border of the forest on his epic morris dance from London to Norwich. Kemp had been the fool in the Lord Chamberlain's Men for whom Shakespeare wrote the parts of Dogberry in *Much Ado About Nothing* and Bottom in *A Midsummer Night's Dream*. In 1598 he was listed alongside Shakespeare and Richard Burbage as one of the five shareholders in the company. By the following year, however, he seems to have fallen out with the playwright and left the company, with one commentator blaming his habit of 'freely improvising his clowning merely to please the worst elements in the crowd'. Before he had become a player Kemp had served in the household of Robert Dudley, the Earl of Leicester. He is believed to have fought in the Low Countries and carried letters back to London for Sir Philip Sidney, delivering them, like Harpo Marx on the rampage, to Lady Leicester, the Earl's wife, instead of to Frances Walsingham, Sir Philip's wife, as intended.

Not long after leaving the Lord Chamberlain's Men, maybe in response to a bet, or possibly for publicity purposes, or perhaps because he had nothing else to do, Kemp embarked upon what he called his Nine Daies Wonder—his morris dance from London to Norwich. Huge crowds followed him much of the way along his route, at a time when morris dancing had come to be viewed as deeply suspect: 'Although the dancers were at one point a force for social cohesion, having been branded as anti-social they became so—disturbing the peace and breaking the law.' In contrast to the perceived passivity of Clare, Kemp mocked and heckled: 'Is it not better to make a foole of the world as I haue done, then to be fooled of the world as you schollers are?' Even an *entertainment* like the Nine Daies Wonder could easily be seen as a serious subversion. Not that it did him much good—if we have a theme at the moment, it's that subversion rarely does. He spent the next few years in poverty and was dead by 1603. The church record read *Kempe—a man*.

If Ken Campbell played the Shakespearean fool as reality instructor, then Will Kemp lived it. Indeed, if the stories are to be believed, he lived it rather too forcefully for Shakespeare's taste, was certainly too independent to be contained in a drama. Appearing as a character in *The Travels of The Three English Brothers*, Kemp is asked what new plays are popular in

London. He replies, 'the old play that Adam and Eve acted in bare action Under the figge tree drawes most of the Gentlemen', and there's something of Campbell's attack on the distinction between performance and reality in that.

What is the lesson of the Nine Daies Wonder? A few months after the event, Kemp published a pamphlet in order, he said, to prove that he had completed the task. The most striking absence in the account is any *reason* for having undertaken the jaunt. Towards the end there is a mention of having 'put out some money to have threefold gaine at my returne', though he also points out that many of the people who were supposed to have paid him would 'not willingly be found'. But at root, the journey seems to have been a performance without a theatre, an outreach programme to 'the worst elements in the crowd', a bit like the Ken Campbell Roadshow. What did the performance demonstrate? The absurdity of status, perhaps. Because if you can be met like a hero for dancing like a fool for eighty miles, what does that say about the heroes? 'Master & men as madmen stop / Life lives by changing places.'

John Clare changed places in the madhouse, fracturing his being across a series of roles. He began telling people he was Byron, or Jack Randall, the prize fighter. Later he added Nelson to his repertoire. It isn't entirely clear what he meant by this identification, how literally it should be taken. In the poem quoted above he had already written, 'A man there is a prisoner there / Locked up from week to week / He's very fond they do declare / To play at hide and seek.' When Clare talked about 'being' Byron I wonder whether this was in a similar vein to David Bowie *being* Ziggy Stardust. Reading *Child Harold*—along with *Don Juan* one of the two long Byronic rewrites he worked on while incarcerated in the asylum at Lippits Hill—I couldn't help thinking of that old saw about Bob Dylan going electric. A rebranding, an attempt to dig himself out of the hole he found himself in as *John Clare— nature poet*, a desperate effort to escape. In the preceding years Clare had tried out a number of pseudonyms—James Gilderoy, Frederic Roberts and Percy Green—in order to pass off as originals the Elizabethan pastiches he kept writing. He had an unfortunate tendency to *admit* the deception, though, and was dissuaded by his publisher from trying to put together a whole volume in that style in 1829. The problem remained the same for Clare, though. By that time he couldn't get anyone to publish a volume of his poetry in *any* style. As John Clare he was already damaged goods, locked into a persona which clearly wasn't working for him.

If *Child Harold* is Bob Dylan going electric, then *Don Juan* is Clare's gangsta rap album:

> Children are fond of sucking sugar candy
> & maids of sausages – larger the better
> Shopmen are fond of good sigars & brandy
> & I of blunt – & if you change the letter
> To C or K it would be quite as handy
> & throw the next away – but I'm your debtor

(Unlike in gangsta rap, *blunt* doesn't refer to a cheap cigar packed with hydroponically grown cannabinoids but is slang for money. Cunt still means cunt). The poem drips with a kind of angry misogyny which seems anathema to the popular idea of Clare as a gentle soul and it's impossible not to read it as an attempt to break out, violently if necessary. But much

111

like rappers today, marketed for their authenticity, Clare was trapped. In the 1820s, when he achieved considerable success (his first volume was shifting six or seven copies for every Wordsworth or Keats that year), he was sold as a *peasant poet*. He was supposed to give the reader a sense of how a son of the soil saw the natural world (much as the hip-hop star is supposed to supply the white suburban listener with an authentic view of black inner city life). To the upper and middle classes, the main consumers of poetry, the peasant was little more than an animal, a lower creature than themselves, so of course his view of nature would be true. He *was* nature. That's what led B.W. Proctor, a contemporary who peddled his own vastly inferior poetry under the name Barry Cornwall, to call Clare 'as simple as a daisy'. At its best this line of thought led eventually to the theory that Clare was a uniquely modern poet who didn't allow anything, any 'higher' thematics, to get in the way of the purest observation. Edmund Blunden, the war poet friend of Sassoon and Robert Graves, said that 'what he saw, that he was'.

The cleverness of that 'saw/was' conjunction perhaps blinded Blunden to the inaccuracy of the idea expressed. Put simply, it isn't true. For one thing, it ignores the polemic in much of the poet's work. The son of an almost illiterate farm labourer, Clare came to adulthood during the early 1800s, when the impact of the ongoing enclosure of common land made it impossible to survive as his forefathers had done. His hatred of Enclosure at every level rings out again and again in his poems. First and most conventionally, because 'Inclosure came & trampled on the grave / Of labour's rights & left the poor a slave'. But equally importantly because of its effect on the landscape:

> There once was lanes in natures freedom dropt
> There once was paths that every valley wound
> Inclosure came & every path was stopt
> Each tyrant fixt his sign where pads was found
> To hint a trespass now who crossd the ground

These lines, from *The Village Minstrel* in Clare's first collection, were excised from later printings without his permission after his patron, Lord Radstock, complained to the publisher that they were too radical. The 'major conceptual creation of enclosure' is *trespass*. No trespass can exist unless someone owns the land. For Clare, this was a lived reality. His response was a growing obsession with hiding. As Adam Phillips points out, 'Clare began to realise that description—defining and evoking through vivid representation—could be complicit with, and even analagous to, certain forms of ownership... To find and to see... was now to use and exploit.' In his 1832 letter to potential subscribers for a new volume of verse, Clare called his poems *Fugitives*. If you had something precious it was best to keep it secret, to keep it hidden.

It's arguable that Clare was a divided personality long before he began to call himself Jack Randall or Lords Byron and Nelson. He was a divided personality exactly because he was a peasant and he was a poet and, socially, these two parts of him were never meant to meet. Ironically, when you think about it, the literary world of his day could never accept Clare as a craftsman, as someone who worked and worked with words until he got them just right. He was a *natural*, who was presumed to do what he did without thinking about it, much as a daisy grows in cow dung. T. S. Eliot, in a famous essay in his collection *The Sacred Wood*, states that 'Poetry is not a turning loose of emotion, but an escape from emotion; it is not the expression of personality, but an escape from personality', and whatever else Clare might have been interested in, escape always rated highly. But at the same time he didn't wish to escape his class, or the landscape he grew up in. He wanted to be an artist *and* a peasant and in doing so he set himself an almost impossible task. 'Very few peasants become artists,' John Berger has written. 'This is not a question of talent, but of opportunity and free time.'

This schism cuts right across Clare's life. The other 'delusion' which has certified Clare down the ages is his apparent belief that he was married to two women at once, Patty Clare née Turner and Mary Joyce. Mary was the daughter of a farmer at Glinton and the pair met at school as early as 1805 or 1806 (it amuses me to report that her father's name was James). As his own father's work became less regular, Clare had to leave the school, but he became reacquainted with Mary a few years later, around 1809, when he was sixteen and she was twelve, and he was working as a labourer and general dogsbody for a publican who lived nearby. They seem to have begun some sort of relationship which lasted until the following summer, before Mary realised that Clare was her social inferior and called things to a close. Or did she? What Clare actually says is that 'she felt her station was above mine *at least I felt that she thought so* ... & fearing to meet a denial I carried it on in my own fancies to every extreme writing songs in her praise & making her mine with every indulgence of the fancy' [my italics]. Fearing rejection, Clare allowed the real relationship to drift while carrying on an idealised version, a manufactured simulacrum through his rhyming. In 'making her mine', he marries her with and through his writing. If that sounds a little overblown then it's worth remembering that Clare stopped attending church on Sundays at an early age, preferring to devote his free day to his versifying. Mary becomes the wife of John Clare the Poet. Poor old Patty, married in 1820, ended up with the role of the wife of John Clare the Peasant.

At around the same time that the short relationship with Mary ended, Clare began to be sent once or twice a week on an errand to a neighbouring village, often having to travel as it became dark. In order to control his fear of the 'midnight morrice dance of hell' he told himself stories: 'I used to imagine tales & mutter them over as I went on making myself the hero…somtimes it was a love story…full of successes as uncommon & out of the way as a romance travelling about in foreign lands & indulging in a variety of adventures till a fair lady was found with a great fortune that made me a gentleman.'

The man who likes to hide and seek embeds Mary Joyce in his poetic practice right from the very start, in the stories he tells himself to ward off fear of the dark, in the rhymes he bounces round his head. Only, more often than not, as time passes, she's embedded in her absence rather than her presence, as if she's just turned her back and walked away. By leaving her out he keeps his precious secret safe.

Even Clare's overtly political and satirical poems can be read in this light. In *The Parish*, written in 1823, Clare sets out to show how the traditional order of village life has been warped and perverted by Enclosure. The root of this perversion lies in the way it encourages the farmer to set himself above and apart from the ordinary villagers: 'That good old fame the farmers earnd of yore / That made as equals not as slaves the poor / That good old fame did in two sparks expire.' The farm as the heart of parish life has been destroyed and what rises in its place 'are built by those whose clownish taste aspires / To hate their farms and ape the country squires'. The year of Clare's romance with Mary, 1809, was also the year when the Act of Inclosure was passed for the land around his home at Helpston. Remember the end of this relationship and the girl who 'felt her station above mine at least I felt that she thought so'. That sentence continues, 'for her parents were farmers & farmers had great pretensions to something then'. Enclosure is an attack on the poor, an attack upon nature and a perversion of human relations. And in the crush of these three impulses, John loses his livelihood, his right to move freely in the land around him and, worst of all, he loses Mary, only finding her again by hiding her, or his idea of her, in and between his words: 'I love thee sweet mary but love thee in fear'.

In his first volume of poetry, Clare taunted the girl to get her attention, like a boy pulling pigtails: 'And mimicking the Gentry's way / Who strives to speak as fine as they? / And minds but every word they say / My Mary'. Once again the poem was dropped from later editions under pressure from patrons who felt it too crude. Once again Clare had not given his permission. And so Mary disappeared twice. In the years after the publication of this first and most commercially successful collection, he dedicated a number of love poems and sonnets to her in successive issues of *The London Magazine*, as if thinking that the story he had told himself to keep his fear and anxiety at bay could still finish with a happy ending, that his poetry had made him a 'gentleman'. When the happy ending didn't materialise, when the magic of his story was broken, his 'nervous fears' intensified. The same terrors he had been prey to as a boy alone on a dark country lane returned to him on a visit to London: 'thin deathlike shadows & goblings with saucer eyes were continually shaping on the darkness from my haunted imagination & when I saw any one of a spare figure in the dark passing or going on by my side my blood has curdled cold at the foolish apprehension of his being a supernatural agent'. As despair and fear began to settle on him, Mary Joyce vanished from his work again. Until he was living at Lippits Hill Lodge, by which time he really needed her.

Clare was ill, on and off, for most of the 1820s and 1830s. After a brief initial success in the early 1820s, before he himself turned thirty, his life had resolved itself into a series of crises, financial and health-related, none of which he emerged from better than he had begun them. His publisher, John Taylor, treated him with a mixture of contempt and condescension. If Clare complained that Taylor was taking a long time over the editing of some poems, Taylor would ignore him. If he wrote again Taylor would blame Clare for submitting poems that needed so much editing. His accounting was non-existent, then inaccurate, just a little dishonest. Along with Clare's other supporters, he liked to tell him how lucky he was to be outside working the land, even though this was something Clare was forced to do—when he could find the work!—to feed his seven children and service his growing debts. 'Every discontented tradesman declares the poor peasants lot to be the happiest in the world & the hardest labour the best exercise,' Clare pointed out, 'yet he would choose any lot than that of a poor man & any labour but that of hard work.'

When his patrons weren't envying his simple country life they were suggesting he would be healthier if he ate and drank less. Many commentators agree that Clare suffered, at various times in his life, from malnutrition. It wasn't a diet he needed, it was more money. By the end of 1831 he owed two years of rent on his cottage in addition to unpaid doctor's bills. Whatever action his benefactors took on his behalf tended to be ill thought out and underfunded to achieve its purpose. In 1832 it was arranged to move Clare and his family to a newly built smallholding at Northborough so that he could benefit directly from his own labours. Unfortunately, no one seems to have thought about the money which might be needed to buy livestock and so on. The family ended up with a small patch of land but nothing to grow or graze there.

It was around this time, with Clare in the deepest despair, that Mary visited him, and although in many respects she was unchanged, she was also greatly transformed: 'that Guardian spirit in the shape of a soul stirring beauty again appeared to me with the very same countenance in which she appeared many years ago'. Clare's muse had become a Muse, a kind of Renaissance magic ascent from the human to the divine: 'these dreams of a beautiful presence a woman deity'. Around this time, Clare composed the poem 'Ive ran the furlongs to thy door'. Mary isn't named but it's clear that this is one of his 'fancies'—a return to his true love. In it he writes, 'Far far from all the world I found / Thy pleasant home and thee / Heaths woods a stretc[h]ing circle round / Hid thee from all but me'.

Everything is always changing in the woods. There's the yearly cycle of growth and death, of course, and the daily iteration of this in new buds, new blossom, new leaves, dead blossom, dead leaves and so on, the tiny, inconsequential shifts which gradually add up to something more, ants moving soil one crumb at a time. But there is another cycle at work too, on a macro scale which escapes human notice as surely as the micro does. This one runs to centuries. Year by year trees grow. Their bark hardens, becomes rutted and gnarly. A limb slowly, gradually decays and falls away. They are struck by lightning or blown from the ground, eaten by beetles, carved by scouts. They morph and stretch and become huge and then they die, rot and gradually mulch down back into the ground. Lippits Hill Lodge lies a little way south and to the west of the trees Doom and Lovely Illuminati, right at the bottom of Hill Wood, and although those trees would have been there when Clare was wandering around, he would not have seen the *same trees*. The landscape might be broadly similar but in another sense, just as real, it is completely different.

I went looking for the site of the little chapel that used to sit in the woods near to Lippits Hill Lodge in Clare's time. It seemed like a perfect place not to find him, as he never went in, choosing to sit outside composing poems in his head. It was abandoned in 1873 when a new church was built nearby, and torn down in 1885 after 'excursionists' had broken in and 'behaved very disgracefully'. I was looking not for a clearing but for its opposite—a patch of growth, an explosion of nature such as would have fed on the light caused by a gap in the trees. I found a crush of brambles, nettles and bracken, some birch and a few thin ash trees, a moss-covered nubbin of brick. The yew tree pressing up nearby convinced me it was the spot, though I had no real evidence. If there was an atmosphere it was of indifference to me, indifference to us all. Nature chews up bricks and spits out tendrils. It has no interest in us, none at all. Clare would like that, the blankness of it.

·

In 1837, John Taylor hit on a new scheme. He had recently published a book by a Dr Matthew Allen called *Essay On the Classification of the Insane*. Allen was one of a new school of psychiatric doctors who believed that 'the grand principle of treatment is, to avoid even the appearance of unnecessary restraint, and to treat [the patients] with a confidence which will almost invariably excite their secret, but proudest endeavours...There is no influence so powerful as the sphere of a moral influence.' Beginning his career at York Asylum, he then moved south, bought some buildings in Epping Forest and started his own business there. Taylor contacted Clare's benefactors and together they raised enough money for Clare to be taken on as a patient at Allen's establishment. He sent a messenger to Northborough with a letter that read: 'The bearer will bring you up to town and take care of you on the way...The medical aid provided near this place will cure you effectually.' The man took him straight to Dr Allen's house, Lippits Hill Lodge, where he was to stay for four years.

At first Clare seemed to think he was enjoying a rest cure—and nobody appears to have bothered to explain the situation to him. He wandered in the forest and wrote some verse about the view. He posted letters home saying he was feeling better. Allen took steps to sort things out permanently. His opinion was that 'if a small annuity could be secured for him, he will very soon recover and remain so'. He made some efforts to raise the capital for such an annuity, but changes in fashion had made Clare less of an attractive cause than he would have been fifteen years before. Money was sent but as it wasn't 'enough' to take him home free from anxiety, it was instead spent on his care at the asylum. Nobody consulted Clare on the use of these funds raised in his name. Nobody told him he was supposed to stay there forever. The inmates of the ayslum were instructed to carry out physical labour in the gardens, the fresh air and exercise supposed to do them good. For Clare, this work was no different to his previous life, except now he wasn't paid. He began to write letters and journal entries with Every Word Capitalised, as if he was hiding acrostic secrets. Still searching for encrypted spells, I tried to unravel them and, in failing, decided only that I'd failed, not that there were none to find. The restraints in his mind loosening, Mary suddenly popped up all over the ground of his poems like mushrooms on damp mornings.

'Mary thou ace of hearts thou muse of song / The pole star of my being and decay'. Is it too ridiculous to think that Clare decided to free himself by his walk home to Mary Joyce not just from the asylum at Lippits Hill but from the whole apparatus of his oppression? That turning his years of wandering the countryside into a purposive trek would somehow bring about the reunion he wanted and, with that, help him shrug off all the problems that had rained down upon him over the last forty-eight years? That in confronting his long-held fear of the dark, by immersing himself in shadows and blackness, the road at night, he would somehow become whole again, complete? This walk—this *escape*—was his attempt to emerge triumphant from a final crisis, his reason for being placed in the forest to begin with.

On the first night of his journey, 'I slept soundly but had a very uneasy dream I thought my first wife lay on my left side & somebody took her away from my side—which made me wake up rather unhappy I thought as I woke somebody said "Mary" but nobody was near—I lay down with my head towards the North to show myself the steering point in the morning.' The steering point by night being the pole star: 'My guide star gilds the north—and shines with Mary'. There was to be no reunion. In 1838, a year after Clare entered Allen's asylum and three years before he escaped, Mary Joyce had died in a fire at the farm in Glinton. But then again, there was no reunion to be had. Clare claimed he didn't believe in magic, though he was haunted by it too, by saucer-eyed goblins and deathlike shadows. Had his plan failed because of this diffidence, his inability to fully commit to the ritual he was trying to carry out? Was it because of his fear? Or was it simply because there was no way it could have succeeded?

Reluctantly, I visited the Suntrap Centre, to the north of the non-existent chapel and just up the road from Lippits Hill Lodge, an education centre which works with local primary schools, bringing kids from Walthamstow and Leyton out to spend a few days in the woods. Back when Allen was running things, this was the site of Fair Mead House, the building in which female patients were housed. Struggling with funding cuts and matters considerably more pressing than dealing with another Clarian, someone still took the time to show me a photocopy of the book the 'Labourer/Poet' was logged in on arrival and the plans of the old house. It was done very kindly, but I knew I wasn't the first. Iain Sinclair had been here too, though he hadn't come in, apparently sharing my aversion to No Entry signs. I was feeling nervous about Sinclair, who had already written a whole volume on Clare's journey and who was, anyway, the first person anyone would think of if you mentioned a book about landscape and poetry and madness and so on. More than anything else I wanted to avoid him, to hide behind a metaphorical bush as he trudged past on one of his poetical tromps. I kept wondering if I had managed it or if that was his breath I could hear, heavy in my ear, coming up behind me. I continued running, crashing further from the path.

My lack of composure was fuelled by Clare himself. I was baffled by him, uncomprehending. *Don Juan* was eye-popping, *Child Harold* an exercise in bewildered pathos. I preferred the poems unpunctuated and misspelt as in the original manuscripts, with the added *doth*s removed. But as a character, as a person, I found it hard to get a grip on the peasant from the Midlands. Rationally, I could see why he did what he did (though that lifetime obsession with a first love went a little over my head), but even while I understood it—both in terms of his specific place in society and the times he lived in—I found his passivity infuriating. He took me back to my hatred of hippies which was, I suppose, a hatred of self. John Clare was furtive, a fugitive, living in fear. I had begun to see the forest as somewhere I could slip free from fear, or even confront it. I'd begun to think I was looking for examples of just this process, of people using the place to vanquish their own timidity, their urge to play it safe. That this was the magic I was seeking, the cure for my own half-hearted problems. Clare's own attempt to do just this had resulted only in swapping one asylum for another.

In John Berger's essay 'Ideal Palace', quoted earlier but not named, the writer talks not only about how hard it is to be a peasant-artist but about the unique qualities of the work created if those difficulties are transcended. 'The computer,' Berger suggests, 'has become the storehouse, the "memory" of modern urban information: in peasant cultures the equivalent storehouse is an oral tradition...the computer supplies, very swiftly, the exact answer to a complex question; the oral tradition supplies an ambiguous answer—sometimes even in the form of a riddle—to a common practical question. Truth as a certainty. Truth as an uncertainty.' That goes some way towards explaining Clare's diffidence while suggesting that a binary approach to Clare the poet and Clare the peasant is flawed. After all, Mary wasn't just the pole star of his being but also his decay. We will return to him. For now, let's take a look at his opposite, the good/bad doctor who was his protector and his jailer, a thoroughly unlovable man for whom, nevertheless, I feel a sneaking affection.

Matthew Allen was a Scottish physician, the son of a Sandemanian evangelist, Christian primitivists who refused to acknowledge the authority of the Church of Scotland or any national church, for that matter. William Godwin, the proto-anarchist and father of Mary Shelley, was trained by a Sandemanian minister. After Calvin had damned ninety-nine in a hundred of mankind, Godwin said, Sandeman had 'contrived a scheme for damning ninety-nine in a hundred of the followers of Calvin'. Clare's attachment to the forest around Epping is entirely coincidental—he was imprisoned there. Allen, on the other hand, chose to come here to set up his clinic—or if you prefer, his business—in 1825, 'as the best adapted of any part of the country about London which I saw'. It's easy to imagine why. The forest offered space, quiet, calm, fresh air, pretty views, yet, only eleven miles from Mansion House, was near enough to London that rich customers—and their visitors—could be there in half a day. If you wanted to make money out of a new take on psychiatry, this was the place to be, not York. If, in addition, you could get books published and even find a relatively famous patient, that was your marketing taken care of. That must have been Allen's thinking, anyway.

Allen's establishment was, according to Jane Welsh Carlyle, who visited in 1831, 'a place where any sane person might be delighted to get admission. The house or rather houses…are…without the smallest appearance of constraint – And the poor creatures are all so happy.' Of course, the appearance of no constraint is not the same as no constraint. Foucault would have had a field day with Allen. Clare wrote to the doctor after his escape to say he could have stayed if it hadn't been for 'servants & stupid keepers who often assumed as much authority over me as if I had been their prisoner'. All the same, compare Clare's treatment in 1837 to Wally Hope's in 1975, or Theo Garman's in 1954, for that matter, and you can say this: at least they didn't kill him. Allen may, on the other hand, have tried to bore Clare to death. He liked a sermon.

Reading the doctor is hard work, a kind of non-stop, endless intellectual hangover. His relentless positivity and wide-eyed bumptiousness are exhausting, as if you are constantly recovering from a party of which you have no recollection. Between 1830 and 1840 he pumped out books—on insanity, yes, but also a series of *Devotional Lectures* on Christian forbearance for families and schools, a course of lectures on Chemical Philosophy, even sermons on *The Redeemer and Redemption Addressed to Jews and all other Denominations*. He didn't exactly go out of his way to be captivating, nor was he what you'd call a scintillating prose stylist. He was also subject to mania, which he describes in his own work on insanity as 'a very exalted state of over-excitation'.

His earliest book on the subject of madness, 1831's *Cases of Insanity, With Medical, Moral and Philosophical Observations and Essays Upon Them, Part 1, Volume 1* begins with a sober statement of intent, to present a series of ordinary case studies: 'I have taken the cases in regular rotation, rather than the common mode, of making a selection of extreme ones, that I might not give a distorted picture of the insane'. Allen manages twelve pages of these case studies before he notes the influence of weather and atmosphere on the insane. The next eighty or so pages are dedicated to this argument, basically that we are all affected by atmosphere, but that the insane lack the 'moral agency' to control the changes that the weather causes in their character. Having done with the weather, Allen moves on to the lunar influence, then the influence of the seasons, then—another twenty-five pages later—'the planetary influence as well'. On page 131 he states that the essay on atmospheric influence actually finished on page 75 and what has followed 'is an appendage which has insensibly increased'. Despite this glimmer of self-knowledge, he devotes the next sixty pages to a discussion of cholera, before returning to the case studies from page 195 until the book's finish only seventeen pages later. The book he wrote for Taylor in 1837 at least sticks more closely to its subject and has some good illustrations of insane heads (I liked the dribbler on page 121 best—shades of American gothic illustrator Edward Gorey). Jane Carlyle's husband, the author and historian Thomas Carlyle, beady-eyed as a crow, described Allen in four words: 'speculative, hopeful, earnest-frothy'.

By the early 1840s, though, Allen's mania had taken a new turn. He had been running his asylum for fifteen years. Perhaps he needed more money. Perhaps he felt he deserved more money. Perhaps it was just what he was like, to be always looking for the Next Big Thing. He hatched a scheme to set up a company to sell a machine which would reproduce hand-carved wooden furniture. In those early days he called it a *pyroglyph*, a name which was eventually dropped for being 'too full of *meaning* (a singular reason for rejecting a word!)'. This sudden change of career demands our attention. As Allen put it, his psychiatric method was built upon 'the effects produced by always treating them [his patients] as rational beings'. But rather than presenting the orderly, moderate, morally controlled environment he hoped for, in the words of Sara Maitland, 'forests, like fairy stories, need to be chaotic—beautiful and savage, useful and wasteful, dangerous and free'. There are no straight lines in a forest, no easy appeal to rationalism, utility or morality.

This was another of my nascent Forest Axioms, one proved repeatedly in those early months as I blundered around, lost. Well, maybe not the morality part, though I was followed one day by a somewhat forlorn Eastern European gentleman who seemed to hope I might have sex with him. And indeed, in Clare's own poems of the time there is more than a hint that behind the front-of-house respectability of Allen's establishment there was quite a lot of wanking and sucking and shagging going on. Allen highlighted the frustration his environment caused him when he wrote of 'the feeling of being excluded from rational society, [which] often presents itself to the mind as a terrible sacrifice to those whose earliest wish was to live in the sphere of intellect and genius'.

How fitting, then, that the doctor fixated on a steam-powered industrial device for turning trees into furniture. Maybe he was scared out there, alone among the lunatics and the beeches, the unthinking rapacity of nature. Maybe he was scared of himself, of what he might do. It's hard to imagine a better metaphor for the reassertion of rationality and utility than the pyroglyph. How fitting, too, that he would eventually reject that *meaningful* name, which made the machine sound magical, which hinted at his desire not just to carve wood but to burn it. Enclosure and the Industrial Revolution go hand in hand, are two facets of the same process, the land being cleared for more intensive farming, displaced peasants moving to the cities (including, of course, the East End of London) to work in new factories. Presented with a choice between trees and chairs, Allen picked the latter. His morality factory would become a furniture factory.

In 1840 the doctor was visited by Alfred Tennyson, still a young man and not yet known as a poet, who was very taken by his establishment and even more taken by stories of his steam-powered carving machine. Desperately looking for a big financial payout so that he could marry the girl he loved, Tennyson invested everything he had in The Patent Decorative Carving and Sculpture Company. He and his mother came to live nearby at Beech Hill House, since knocked down and replaced with the kind of footballer's mansion the area is renowned for. While the young poet was away, Allen persuaded Tennyson's mother to invest everything *she* had in his scheme, as well. Over the next few months the Tennyson sisters were brought in too. Allen was particularly keen to get the richest brother, Frederick, to invest, but Frederick, being already wealthy, had no need for get-rich-quick schemes. Only those members of the family who couldn't afford it were tripping over themselves to put their paltry savings in.

Allen's background wasn't quite as clean and morally upstanding as he may have let people believe after his move to London. He had twice been imprisoned for debt and overseen the financial collapse of a soda water factory in Leith. He had started at York Asylum not as a medical doctor but in the pharmacy and, in fact, his medical qualification was granted a few years later as a formality because of his work as a chemist. In 1841, Allen claimed that he had been tricked by an agent (possibly supernatural?) and lost a large proportion of the original investment. He sent a series of letters to Frederick appealing for funds: 'it was the decided opinion of my Bankers and Solicitors that in twelve months your share would be worth Ten thousand Pounds, and that in five years it ought to give you that yearly'. As late as May 1842, Allen was telling Alfred to 'have faith and all things will be more than well'. Between them, his family had invested eight thousand pounds. Allen's business dealings were chaotic and the machine itself never seems to have materialised. Perhaps it never existed.

By 1843 the whole enterprise had imploded and Allen was writing to Frederick once again, begging for assistance: 'I have done all that is possible for man to do to save your family, and I have *utterly ruined myself* in the attempt... Every stick and stave is to be sold to pay A. T. *this day* – and yet people boast! I ail! and I suffer! and I die! – *Come up!* – you must save something.' Assistance was not forthcoming. Somehow, though, the asylum's furniture was not sold. Tennyson was bankrupted, but Allen's core business limped on. Early in 1845 he had a massive heart attack and dropped dead. By chance, the doctor-entrepreneur had given Tennyson a life insurance policy back in 1840— presumably to protect Alfred against the premature loss of this scalpel-sharp Scots business brain—so the poet now got his money back. Allen's wife continued to run the asylum on her own until the early 1860s. Tennyson channelled his anger and hatred of Allen into two long poem sequences, *Sea Dream* and his overwrought psychodrama *Maud*.

Speaking of overwrought psychodramas, I went to see the collaborative film about John Clare that the artist Andrew Kötting made with Iain Sinclair, my way of turning to face him as he lumbered up behind. It was being shown as an installation in a former church hall in a park in south London, very dark, cold and damp. My son thought it was a horror movie. There was plenty of lurching around, lots of filming from just behind someone's shoulder and whoever put together the soundtrack obviously had a great deal of fun making deep, off-kilter droning noises. Toby Jones appeared as a teddy bear-shaped Clare, Andrew Kötting himself featured as a Straw Bear—looking like a cross between a wicker man and a clumsy addition to a Malawian dance ritual—and Iain Sinclair played Iain Sinclair in a billy goat mask, I guess. Alan Moore, the former comic book genius turned Northamptonshire Archmage of Weird, also made a guest appearance sitting on a bench and reading one of the poems, rather beautifully.

It was Sinclair I was watching, though, this big, galumphing, slope-shouldered presence in his little mask. In the TV show I'd seen about dogging in the forest, the director had got round the participants' need for anonymity by disguising them with animal masks—an owl, a hawk, a fox, a pig and so on. It was the masterstroke of the whole series, making the most banal statements—and let's be honest here, *all* the statements were banal—sinister and alien, creepy, dripping with discoloured spunk. Now here was Sinclair tracking John Clare through those same woods in his billy goat mask, and that memory of the TV doggers made it look to me like he was eager either to roger Clare or to watch him rogering someone else. Probably Mary-Joyce-as-goddess, in a basque and ripped stockings, her duck mask skew-whiff as John pumped away, his breeches round his knees. Iain, I thought, sniggering to myself, would really get off on that—and as soon as it occurred to me I felt somehow stronger, more sure of myself for having lampooned this particular god.

It was only later that I noticed another, smaller figure had invaded my imagined satire, standing behind Sinclair, jumping up and down to get a view, too scared to push the alpha male out of the way. It took me some time to make out the face and even longer to recognise it. That figure, of course, was me.

CHAPTER 6

Dial House | Countesthorpe Community College

The first time I saw Penny Rimbaud he was dressed in straight-legged trousers of a dusty burgundy colour, hiking boots, a black vest and, atop his long, fine, grey-blond hair, a red beret with a red star the size of a child's fist embroidered on the front. Penny is in his early seventies but possessed, appropriately considering his adopted surname, of a certain ageless *je ne sais quoi*. Even when cutting the hedge.

I realised where I was going as soon as Rimbaud emailed me directions. I was retracing the route of my attempt to find tragedy on the train line from Epping to Ongar. I rode my bike up the same hill from North Weald to the same strange fort, which I now knew to be a derelict radio station, then on down the same track, past the same No Entry sign which had previously halted me. Although I was invited on this occasion, I wasn't invited by the owner of the sign and I still tensed as I trundled past the ugly new-build Olde Worlde house and through one of those filthy concrete farmyards which remind you definitively that agriculture is an industrial business. I waited, shoulders hunched, for the bark of a sheepdog even though there were no sheep but me. Then I turned the corner, trundled through another expanse of muddy cement and clambered off my bike at the start of the footpath. Ahead of me was a man with a colossal electric hedge trimmer, not some regular domestic clipper but a gigantic, elongated, industrial arm of a thing. It was, after all, a big hedge.

When I greeted Penny he stopped strimming and smiled, essaying a lazy wave, then began gathering in orange electrical flex. We ducked through a gate adorned with brightly coloured oriental flags, stepped over the mosaic in the concrete at our feet and walked round the side of a large old house, all wooden wings and tumbled perspectives so that it was hard at first sight to get a sense of what it was, how big it was or how it fitted together (it turned out later that this was a theme continued on the inside). Penny led me into a garden which I can only describe as beautiful. The lawn was short and very green, flowers flooded up from the beds, bees moved calmly through the warmth of the air. It felt very easy to be there. All my nerves about the meeting ebbed away into a kind of sunstruck haze. Penny fetched us cups of Earl Grey tea and called Gee so the three of us could talk.

Gee's a similar age to Penny, with grey hair down past her shoulders, a brown face, strong hands. Like Penny, she knows something about everything and isn't afraid to tell you, though their styles of delivery are very different. He circles in on a problem as if from a great distance, orbiting closer and closer. She is quietly direct, full of good humour but not to be messed with. Before I knew what was going on the two of them were having a dispute about whether the development of the internet signalled a definitive cultural break with the past or a seamless evolution. They carried on the conversation like a married couple bickering over who had forgotten to defrost the freezer. They were not a married couple, though, and had not been a couple at all in any romantic or sexual sense since the early 1970s.

It was quite dizzying but at the same time strangely relaxing, sitting there listening to them, a thrush calling from a tree, Gee leaping up from her chair to point out a passing steam train up on the track which ran along the bottom of the garden. Back in the late 1970s and early 1980s, when the line was still part of the London tube network, the drivers often used to stop on the bridge we could see through the trees so that the young punks who had travelled from all over the country could get out and scramble down the embankment. They came here to meet Penny Rimbaud, Gee Vaucher, Steve Ignorant, Eve Libertine, Joy De Vivre, Pete Wright, Andy Palmer and Phil Free. They came to Dial House to meet Crass.

You probably have to be of a certain age to remember Crass, somewhere between ten and twenty-five in or around 1980. Arguably the most radical, polemical and politically engaged band to emerge from punk, and undisputed leaders of the second wave of that movement, they have largely been painted out of successive revivals of safety pin and bondage trouser nostalgia because of their refusal to *fit*, something which they saw as the point of punk even as the rest of this supposed explosion of non-conformism found new and increasingly craven ways to belong. They can lay a reasonable claim to have introduced notions of radical feminism, vegetarianism, anti-Christianity, anti-authoritarianism and pacifism, if not into the cultural mainstream of early 1980s Britain, then certainly into the lives of many young men and women who were trying to escape its crushing grip. I can think of no better way to extol their virtues than by pointing out that their first album was slagged off by both Garry Bushell (in *Sounds*) and Tony Parsons (in *NME*), the latter describing it as 'a nasty, worthless little record'. With enemies like that, who needs friends? When I met Penny Rimbaud for the first time he had recently returned from a punk festival in Blackpool where he and a bunch of jazz musicians had presented what he described as 'a Taoist reinterpretation' of the most out-there, avant-garde record in the Crass catalogue, *Yes Sir, I Will*. He seemed both mildly surprised and somewhat disappointed that he hadn't been bottled off the stage.

The mention of Taoism was an early clue in the conversation as to the direction of Penny's gaze at that time. Born Jeremy Ratter, a man who reinvents himself at regular intervals (and also one who takes occasional pleasure in outright lies), that afternoon Rimbaud was an ageing Zen master. I'd say that we spent a long time discussing quantum mechanics, but that would suggest I was capable of discussing quantum mechanics, or even classical physics for that matter. I listened and nodded and tried to look as if I were following. At one point he told me of some time he'd spent as an artist-in-residence on this university campus or that, where he had passed his days in the physics block, trying to get the students and professors to agree that it was possible under quantum theory that a person's neurons could be outside of them. Or something like that. My brain had cast itself as Schrödinger's cat—as long as I didn't say anything it was neither alive nor dead. At some point we were joined by Gordon, a large, good-humoured man with an evangelical love of vaping. I think he may be Gee's brother, but I could be wrong. Penny stuck to his roll-ups and made a seamless transition from theory to anecdote.

He talked very calmly, concentrating very hard on what he was saying, often closing his eyes. Although he took what he said seriously, he didn't seem to take himself too seriously. All the same, it's fair to say that he liked talking. He said some pretty mind-blowing things, both then and when I met him again, things which seemed to me ridiculous or borderline laughable. But he said them in a way which stopped me laughing at them, somehow. Because he was being honest. He wasn't tailoring what he said to what he thought I wanted to hear. He wasn't trying to avoid saying anything which might open him up to ridicule, nor saying crazy things to impress me. He was also unfailingly self-critical, so you knew he wasn't saying these things as the kind of undigested bullshit which people—particularly musicians who have spent years subjected to soft-soap interviewing and fawning fans—tend to come out with. Although, to these ears at least, some of it was still bullshit.

I think it's fair to say that it's largely because of Penny Rimbaud that Wally Hope is remembered and, in certain circles, revered today. That wasn't why I'd come to see him, exactly, but in terms of the path we've been wandering down I should deal with it first, not least because it became a founding myth for Crass too. And anyway, it ties in with my real reason for coming to Dial House. Part of that was to do with how Penny came to tell the story of Wally Hope, and what I could glean from that about fear and bravery. Part of it was Dial House itself. Penny has lived here, right up at the very northern tip of our expanded, thirteenth-century ghost forest, coming on for fifty years. When he arrived in the late

1960s the building was completely derelict and the farmer who owned it was thinking of bulldozing it. He didn't even pay rent for the first three years.

For a while Penny lived there with a couple of other teachers from Loughton College of Further Education, where he taught art, but when he left that job he kicked them out too. Or rather, he told them he wanted to make the house into a kind of commune and that saw them off. Which probably would've been my reaction. But Penny had watched a film, *The Inn of the Sixth Happiness*, and become enamoured of the idea of a place where anyone could come and stay and be fed for a night in return for telling a story. To do it in Essex, though, and not some exotic location in the Far East, was what made it experimental. As Penny remembers it, 'I wanted to take the locks off the doors to see what would happen.'

He never really advertised or told anyone, but pretty soon people were moving in. 'None of them were paying anything or doing very much,' he recalls. 'I was working on the coal and then coming back every night knackered and they'd all had a nice day sitting around chatting. It seemed a bit unreasonable. But I wasn't quite sure what you were meant to do about that! Lots of stories, not much help!' Gee and Penny were a couple back then, having met at Dagenham School of Art. She had been living in another house locally, but at some point she moved in too. Predictably, the house became a magnet for the local youth, a little bubble of autonomy on the edge of the forest. That was how Wally Hope turned up. He heard about it from local kids when he was visiting the Hatfields in Ongar and came to check it out. It was here that he formulated the idea for the Stonehenge Festival. Gee and Penny helped with posters, flyers and artwork. I've already told you what happened after that. What I haven't told you is what happened after what happened.

There are endless other places you can read all this stuff, or watch or listen to all this stuff—both the story of Wally Hope and Penny's place in it. I'm not going to retell what's already been well told. The quick version is that Penny, devastated at the death of his close friend, researched a book about Wally's demise and came to believe that he hadn't merely been killed by *the system* in some broad, impersonal, crushing kind of a way, but was assassinated. There was a syringe mark on his arm, massive inconsistencies at the inquest and so on. But in researching the book, Penny became scared, paranoid, lonely, was visited by people who threatened to kill him too, out there in the country, in a house with no locks on the doors, and once the fear has blown its way in it's hard to get it out. This crisis was ended by his burning the manuscript, titled *Homage to Catatonia*, out in the garden, a bonfire and two years' work gone. Or that's how he told it—one of the ways he told it, anyway. Darkness closing in, the victory of fear over truth. Except that shortly after that, the young man who went on to become Steve Ignorant turned up at the house, and the early thrashings-around of punk began in London and what became the band Crass started to develop. Penny sat down at the drum kit as well as the typewriter and poured all his anger and frustration, but also all his love and hope, into that. Punk as a Dionysian ecstasy flooding out of the city to its forest home. Catharsis. Making whole, the purpose of tragedy.

Crass released their first album, *The Feeding of the 5000*, in 1978, including a two-minute silence where the track 'Asylum' should have been, retitled 'The Sound Of Free

Speech' after workers at the pressing plant refused to manufacture the record with it on. It was released a year or so later, now called 'Reality Asylum', the first single on the band's own Crass Records, and provoked the first visit by Special Branch to Dial House, where the band were cautioned and questioned but not, eventually, charged (that came later, for another record).

Listening to it now it's still exhilarating, a shocking record, basically a section of a long poem written by Penny and performed by fellow band member Eve Libertine who, in a strict, clipped voice, takes Jesus to task for starting a pathetic, sex-fearing, misogynistic male death cult which reached its apotheosis at Hiroshima and in the death camps of Nazi Germany. If someone were to release this record now I'm sure it would fall foul of some law or other regarding religious tolerance. But in the late 1970s, when almost 12 per cent of the population still attended church regularly (the figure had halved by 2005) and two thirds self-defined as Christian (down to half by 2008), this was about as provocative as you could get. Johnny Rotten may have called himself *the Antichrist* but Crass went somewhere beyond empty gesture into the realm of polemic: 'He shares nothing, this Christ / Sterile, impotent, fuck-love prophet of death / He is the ultimate pornographer.' What makes it work is Eve's voice. She's not affecting a cockney accent or trying to sound *punk*. There is something very clean and Home Counties about the way she speaks, something very controlled, and it's that control which gives the words their power, both through the sense it gives of anger barely restrained and the contrast between this middle-class, Middle England voice and what it says.

In a different way the band played the same trick on their

first single, 'Do They Owe Us A Living?'. This is, on the surface, a standard three-chord punk tune, all distorted guitars, Penny's paradiddling military drum style and Steve Ignorant's East End magma flow of barely scanning lyrics. But I remember it being played to me when I was fourteen or fifteen and the shock of that chorus, which turned upside down all my expectations and understandings of how the world worked and was supposed to work, a call followed by a screamed response, the pressing of a very big reset button: 'Do they owe us a living? / Course they do, course they do / Owe us a living? Of course they fucking do'. No wonder they were slagged off by Parsons and Bushell, with their reductive idea of punk as an expression of working-class sentimentalism; no wonder they called them hippie-punks and lambasted them for growing their own vegetables. I hope that when Tony Parsons finally passes over to the other side he will find a pair of headphones welded over his ears and that they will play 'Shaved Women' to him on repeat for all eternity. That would be too kind for Bushell.

You could pick almost any song from Crass's back catalogue and there would be something to shock you, to make you squirt tea through your nose with laughter or disgust, to make you reassess your assumptions, or theirs, to smile at their naivety and then wonder if it's actually your own. They released five albums in little over five years and every one of them topped the Indie charts (a considerably more important marker of commercial success at the time than it is now): *Feeding of the 5000* was followed by *Stations of the Cross*, then *Penis Envy*, on which Joy De Vivre and Eve Libertine took centre stage as vocalists and emphasised the feminist message of the music, then *Christ – The Album* and finally, the utterly avant-garde, lacerating, reasonably unlistenable *Yes Sir, I Will*. Each record was given increased polemical impact and visual identity by Gee's remarkable covers.

In between releasing albums the band were discussed in Parliament (for their anti-war singles, 'Sheep Farming In The Falklands' and 'How Does It Feel To Be The Mother of 1000 Dead?') and were prosecuted for obscenity (bizarrely for 'Bata Motel' from *Penis Envy*, a record which was banned by HMV and targeted by a star of 1980s moral posturing, the Chief Constable of Greater Manchester Police, James Anderton, who believed that God was using him to achieve His ends). They also tricked the CIA when, using speeches recorded from the TV and radio, they fabricated a tape of a phone conversation between Thatcher and Reagan which they then sent out to newspapers in an attempt to destabilise the British Prime Minister before the 1983 General Election. Nothing was heard of the tape for six months until the US State Department got hold of it and accused the KGB of its fabrication. Only an investigation by the *Observer*, which successfully traced the tape back to Dial House, stopped the situation from escalating any further. The band also helped fund the brief existence of an anarchist centre in London (despite Penny claiming that if you'd asked them about Bakunin they would have assumed it was a make of vodka), and were intimately involved in the promotion of the Stop the City events of 1983 and 1984, now widely seen as the prototype for Reclaim the Streets in the 1990s and—perhaps more directly—the Occupy movement in the new century.

On top of this they started Crass Records, which brought through bands like Flux of Pink Indians (whose Derek Birkett went on to start the One Little Indian record label) and Björk's first band, Kukl, and also established John Loder's Southern recording and distribution empire. Recently Björk reminisced about her own visit to Dial House: 'They were way more radical politically than us ... They had a huge antenna that some crazy electronics guy had made and they were literally hacking into Margaret Thatcher's speeches on BBC radio and creating anarchy.' Appropriately, the band finally split in 1984, on the way home to Dial House from a benefit gig for striking miners in Aberdare (a doubly poignant way to go out, as both Gee and Penny have talked about the role that the Aberfan Disaster of 1966 had in their political awakening).

You may be wondering where the story of the telling of the story of Wally Hope comes into all this. First there was *Homage to Catatonia*, burnt on a fire in 1977, a kind of lost urtext. Next there was a booklet entitled 'Last of the Hippies – An Hysterical Romance' (yes, another romance), which the band included as an insert with *Christ – The Album*. In some ways this is the definitive text. If anyone ever talks to you about Wally Hope, this is what they will be reciting, almost word for word. I think it's true to say that before 1983 no one gave a stuff about Wally. Now there seems to be a whole alternative culture devoted to his sainthood. For Penny, though, the story functioned as an opportunity to expand upon the Crass world view of our complicity in our own oppression, as well as trashing the familiar narrative of the definitive break between the hippie and punk movements and instead turning the distinction enantiodromic: 'this time round they didn't call us "hippies", they called us "punks".' However, Penny still bottled out of the claim that Wally's death was a state-sponsored murder, choosing to present it as a suicide brought on by everything he had been subjected to. As he put it to me, 'I really think I *used* the story in "Last of the Hippies", in quite a cynical way. I used the story of Wally to promote the ideology of Crass. And that was not right.'

I'm not sure I'm entirely comfortable talking about using other people's stories. What is biography if not the telling of another person's tale? We're wandering in a forest where every tree is someone else's life. What Penny was hinting at is that, within the narrative he was building of self-oppression and silent assent—the key theme of 'Last of the Hippies'—Wally's suicide-by-overdose was more powerful (and perhaps more generally credible) than a screed about secret state terror. The first version of Wally's story was burnt in fear. The second version was flawed because it was used to promote the outlook of the very collective which had given Penny the strength to write it, as if a Greek chorus had taken control of the play it was supposed only to provide commentary on. So he tried again. In the 1990s, when he was asked to write a Crass memoir, Penny instead used the opportunity to set the record straight as far as he could, which is why most of the 'biography' of *Shibboleth: My Revolting Life* is in fact concerned with the life and death of Phil Russell aka Wally Hope.

Although great chunks of it are taken from 'Last of the Hippies', with the collective 'we' of Crass returned to an 'I', this time Penny says pretty clearly that he thinks the founder of the Stonehenge Festival was murdered. Incidentally, shortly after the release of *Shibboleth*, Penny was himself knocked off his bike and hospitalised, allegedly by Special Branch: 'Maybe it wasn't an accident. Maybe it was an attempt to put me out of action. Maybe, maybe…' But he still wasn't done with it.

When 'Last of the Hippies' was reissued and he wrote a new introduction to it, he said that one of the key ideas expressed in the booklet—a key idea for him in the formation of Crass and one of Wally's own beliefs—now embarrassed him: 'The fanciful dream…that rock 'n' roll could act as some kind of radical world-changing force was archetypal of a form of inverted racism…I should've known better, but in 1982 I didn't.'

This is one of the lovable things about Penny Rimbaud, a streak of contrarianism so strong that even if he called it that he would immediately have to deny it: 'If I set something up it's almost certain that I will then go into the process of attempting to undermine it. Because that's a way of testing it. I'll play the devil's advocate to test my own thought. So what appears to be a contradiction is simply looking at the same thing in a different way. And I don't feel any contradiction within myself. I perfectly respect that other people might do. And yes I do, I change my tack, and have done for most of my life.'

By the time he told me all this it was autumn. Not because he had taken that long over articulating it, but because we'd reached the end of that first sunny afternoon with Penny agreeing to be interviewed by me. When I told him I'd probably only need an hour or so he shook his head and explained that if he did something he did it properly. I should come back to Dial House one Monday afternoon, we could put in four or five hours, then I could stay the night and we would finish in the morning. Which, allowing for a minor dental emergency, is more or less what we did.

The rain had been coming down since dawn that day, not some minor mizzle but gallons of the stuff, and I'd opted to drive, a long, slow trundle along congested forest roads, each driver's nose pressed up to a steamy windscreen, even the green above me black in the gloom. Gee came to the door— to the second door. I'd tried what I thought was the front of the house, and when it didn't instantly open had sidled round the back to the kitchen. I ducked in to find another guest baking bread for dinner. The baking of bread is a very important part of life at Dial House, and Gee was making encouraging noises as Dominic worked on his very first loaf. Penny was in their living room having a lie-down, listening to a Miles Davis recording from the mid-1950s. The room was small and cosy, stuffed with books, heated by a wood-burning stove. Dial House is basically a warren of small, slightly wonky rooms, one after another, put together with added rickety staircases. Some are cosy, some less so. The less cosy ones smell a little of damp, at least they do when it's pissing down outside. There's a lot of varnished wood, tiles, art. It feels like a getaway, only they've been getting away for fifty years. Dial House has been called a Charleston for the punk generation, and that's not a totally ridiculous way for you to imagine it, only living and breathing, sidelong and not completely up itself, utterly lacking in snobbishness. Once again I felt that same calm descend upon me, a kind of giddy relaxation completely alien to me. I was home and at the same time on holiday from life.

Home in time, I suppose I mean. Or back in time, maybe. 1980 is as far from us now as the end of the Second World War was in 1980. And believe me, in 1980 that seemed a long time ago. The narrative told about the 1980s is always yuppies with shoulder pads and big mobile phones and Thatcher and financial deregulation and the ideological victory of neo-liberalism and so on, but that wasn't actually the decade I inhabited at all. Mine was the 1980s of supporting the miners and then the print workers, of anti-apartheid and anti-racism, of feminism and the first stirrings of political correctness, of vegetarianism and animal rights issues. Some of this came from my family. My parents were (and remain) nonconformist, liberal left-wingers, Labour Party supporters living in a Conservative safe seat on the edge of Leicester (our MP was Nigel Lawson, in Fat Chancellor mode). My sister was a flying picket at the pits and later at Wapping. We would joke about our telephone being tapped until, some years later, we found out that the clicks and echoes we heard probably meant that it was. Some of it came from my friends. Two of the people I most admired and aspired to be like when I was fourteen were fans of Crass and related bands. They wore raggedy army surplus and brand-free black baseball boots, spiked their hair with soap and experimented with a macrobiotic diet that gave them wind of a frighteningly intense aroma. *Fans* is the wrong word, really. Despite the Crass ethos of personal self-determination, as a teenage follower of what became known as anarcho-punk you didn't just favour a certain type of music and polemic, you bought into a whole world view.

The third and perhaps key source of politics in my life was school. Countesthorpe Community College was conceived at the end of the 1960s and opened in 1972, around the time Wally Hope and Penny Rimbaud were mapping out ideas for the first Stonehenge Festival. Seen by some as the apotheosis of the comprehensive system it was quickly jumped upon by others as its nadir. I'm not an educationalist, but my understanding is that the school was supposed to be progressive, open and non-hierarchical, a place where teachers and students, adults and young people, all learnt about learning together. Influenced by private institutions like Dartington (which Margaret Epstein complained had only been good for teaching her daughter, Peggy Jean, to smoke) and Summerhill, it was an experiment in education.

When I tell my children about my schooldays I can see them scrutinising me carefully to be sure I'm not having them on. I tend to avoid talking about the Team system and open-plan layout; the Moots, which could be called by any member of the school community at which every individual—student, teacher or management—had an equal vote; the circular design of the building with the modern sculpture at its heart. Instead I focus on the things that really get them going: that we didn't wear uniform, called teachers by their first name, had no detention system, didn't have to go to lessons and could tell the staff to fuck off if we so wished. There was no staffroom and—perhaps most shocking of all now—a smoker's coffee bar where anyone could puff away, regardless of age. The smell of instant Nescafé and cigarettes still takes me back there, our hair a riot of dye and spikes, our clothes ill-fitting ragwear. As I say, apotheosis or nadir, take your pick. I believe, though, quite strongly as it happens, that because our teachers couldn't rely on the usual vestiges of authority—and on punishment—they had to concentrate instead on being *really good teachers*. Shouting wouldn't do them any favours so they tried to be funny, charismatic, interesting, clever, occasionally sly. The ones who weren't didn't last long. I still remember poor Dave who, straight out of teacher training college, took us for physics and thought that an interest in Everything But The Girl should be enough to divert us from igniting gas taps.

It's probably fair to say that a school like Countesthorpe also attracts a certain type of teacher, and I know that some of the original members of staff felt that, as the 1970s turned into the 1980s, the school was hijacked by what the *Daily Mail* might call *political extremists*. It's more accurate to say that it was a polarising time, that teachers felt threatened and under fire from the Thatcher government and that this led to a hardening of positions on both sides. In addition, you might be tempted to question the wisdom of opening such a school in such a thoroughly Tory area, or have taken a moment to assess the politics of the local media, where the *Leicester Mercury* was to make hay for a number of years with shrill denunciations of my alma mater.

All the same, the politics of some of the staff was pretty hard left, or was reputed to be. There was a rumour that one of the teachers was not only openly lesbian (much more rarely acknowledged then, especially with Clause 28—a law which banned the 'promotion' of homosexuality in schools— approaching the statute books) but, more impressively, a Maoist. Another member of staff was quite blasé about being an anarchist. He took us out in the school minibus on a day trip which involved chasing the Education Secretary, Sir Keith Joseph, round town. I still remember being bundled away by Joseph's security detail having scaled the fence of someone else's school and charged him as he emerged from his tour of the premises. Due to a dispute about pay and class numbers, I think, the whole of the staff worked to rule for most of the time I went to Countesthorpe (1983–7), vacating the premises at lunchtime and leaving our supervision to a fellow student and enthusiastic advocate of acid who, not unconnectedly, had been expelled from Lutterworth Grammar and who we now taunted for being a scab. It sounds like I'm trying to paint it all in a bad light, that I'm playing along with the idea that my school was an educational hellhole. I'm not. It was *brilliant*.

It was only when I was going to bed at Dial House, when Penny had finally run out of energy around one in the morning and agreed to leave it till the next day, that it struck me. Penny has a kind of outhouse, I think referred to as his Shed, in which he sleeps and works. It has big windows in two walls, a large desk, a wood burner, a bed. Take some nice photos of it and you've got a fantastic spread of hut porn. So when we finished he led me to the main house to show me where I was sleeping. As we stood at the bottom of the stairs he whispered to me, suddenly insistent, that I mustn't wake Gee up, that she did not like to be woken up. And I was back, suddenly, aged seventeen, creeping down the stairs from my friend's attic room, stoned and a little drunk, past his mum's bedroom—herself a former teacher from the college—right by the front door. A woman who must not, *at any cost*, be woken up. Penny and Gee reminded me of her. They reminded me of my teachers. The smell in their kitchen summoned up other vegetarian kitchens in the 1980s, where a pot of tahini always lay open, a knife discarded next to it smeared in brown gloop. I recognised the confidence in their own position, even as and when they criticised it. I knew without even thinking about it that these were good, trustworthy people trying to move through their lives in an honourable way.

There was a time in the early 1980s, hard as it is to imagine now, when we thought we would win. We really thought we would win. I can remember us all sniggering at the mother of another school friend who would get drunk at her parties and tell us, incessantly, that *we nearly made it in the sixties*. But looking back now I can say we felt the same way. We thought we could win. I keep saying it in the hope it sounds a little less embarrassing. And there was a time—around 1984, funnily enough, as if the significance of that date was lost on none of us on either side—when that thought seemed to be shattered, when it began to become clear that we had in fact lost, so that when I left home to go to university in the very highest bastion of privilege, my politics were already somewhat tattered and apologetic. And even if they weren't, a year-long campaign to try to get our college to disinvest from South Africa, which ended with us sitting in suits and being patronised—as if we were trying hard in a particularly difficult tutorial—before being told to forget it, that awful meeting and the interminable, bungled build-up to it, killed any last real interest I had in participating, in trying to change anything. Because, more than anything else, everyone had conspired to make it so boring. Although that's not quite true either.

It's also the case that sometimes it's easier to lose, to give up, particularly if in plenty of ways you benefit from that loss. There was something of me summed up in that loss and my acceptance of that loss, of the supposed lefty who has never been in a trade union and started his own business which, in effect, bought up and exploited other people's copyrights. I couldn't think it through at the time, though I lay awake for a long while in my room, afraid to go downstairs again to use the toilet in case I woke Gee. It was too close, maybe, or too far away. The tree beyond the uncurtained window was a black shape, a void within the void of sky, a spreading pool of spilt ink, of bad faith. My pulse thudded in my eardrums and it seemed impossible, somehow, that dawn would ever come. Though of course it did.

We continued with our discussions around half nine, after coffee and more home-made bread. The rain had finally stopped and the cat which came to sit on my lap again was dry now. In the space of almost five hours of recorded interview we covered everything and yet nothing too. We talked about the time Penny spent teaching art at Loughton in the 1960s. We talked about the avant-garde band EXIT, about the ICES festival at the Roundhouse in 1972 that he and Gee helped organise. We chatted about working on the local coal cart and about Fluxus, about experiments with shamanism and magic on nearby Goat Hill. Penny told me about the time they performed one of Anthony McCall's fire pieces at the nearby North Weald aerodrome. How he'd nearly managed to get a TV programme made in which he, Gerry Adams and Arthur Scargill discussed the events of the early 1980s. He also told me bits and pieces about what he'd been doing in the thirty years since the break-up of Crass. His summer spent as a volunteer lifeguard at Summerhill school's swimming pool. Trips to northern India and Africa. An interest in mountaineering. The ongoing engagement with Zen, which ran, on and off, from when he was fourteen until the present day. The parallel—though possibly contradictory—fascination with Existentialism, abundantly evident in the final exhortation of 'Last of the Hippies' that 'THERE IS NO AUTHORITY BUT OUR OWN'. His lifelong pursuit of 'truth and beauty', the fact that he could say those words without having to ironise them part of what I already loved about him. The battle to save Dial

House from the developers, which we'll come back to later. A kind of exile and then return.

But there was one thing I didn't ask him that night, or the following day, though I only regretted it much later. I didn't ask him how they'd coped when the band and the world it was supposed to presage had both collapsed; when Thatcher and her cohorts had definitively won and the rest of us were crushed, when everything we thought we believed in was broken and kicked into the dust. I didn't ask him how come his back was still so straight and his world view so untarnished. I didn't ask him why he was so serene. In all honesty, I didn't ask him anything of much interest. But when I was listening back through the tapes I realised he'd explained it to me, in a funny kind of way, without me having to enquire. He hadn't retained his serenity at all. He'd discovered it.

Three or four years earlier, Penny had found himself in his own dark forest, both physically and emotionally. 'A relationship I was in had fallen apart, which had left me feeling pretty devastated. Then, on top of that, I had to deal with numerous health issues, notably a cancer diagnosis. I really didn't think I was going to survive. Either I'd die a natural death or I'd do myself in. Yes, total self-indulgence, but then, after months of misery, I realised that it wasn't the loss of love from another that was the problem, but the lack of love for myself. I was so framed within conceit that I was unable to break through my own layers.'

He thought for a moment, licked at the edge of another roll-up, looked towards the ceiling. 'I'm not sure I put that entirely well. I had a big operation which broke the chain. I didn't think I was going to survive the operation. They thought it was an extension of the cancer in my jaw. It wasn't, it was just a cyst when they got to it. Anyway, when I came out of the illness it was like it had broken all the chain of grief. The angst had gone. And I just became…Everything that I'd learnt when I was fourteen and that so excited me about Zen, seemed to have come to me. Without any effort. It was like something had just happened.' He paused again, smiled. 'Which is how it happens.'

CHAPTER 7

The Green Man | Claremont Road

I was at the Green Man Interchange, a mile or so from my house in Walthamstow, down towards the bottom of the forest, and I was taking photographs of rubbish—empty paracetamol packets, beer cans and bottles, chocolate bar wrappers, the myriad plastic bags which blew, tattered, from this branch or that of the scraggy little birches, like the abandoned standards of some medieval battle. I'm not sure why I was taking pictures of rubbish other than that it was there. A few joggers thumped past, women pushing buggies, a couple of cycle commuters, the fluorescence of their yellow jackets making the sky a more powdery blue. I'd passed by this spot pretty regularly over the last fifteen years, but, perhaps because of that very passing, never noticed quite how scruffy and threadbare it was. Up above me, beyond the palm trees planted along its perimeter to make it more picturesque, a complete ring of traffic was making a slow and painful circuit, changing and unchanged.

It's a long way from Dial House to the Interchange. They lie at opposite ends of the forest, both geographically and in terms of spirit, one out in the country and the other hunched in the city, one bucolic, the other ugly, one expansive, the other a bottleneck. They remain, though, facets of the same beast, our Essex/London goddess crashed to earth and marked out on the ground by clumps of trees. Nothing here is exactly what it seems. They call it an interchange but to the rest of us it's a roundabout. It was named after The Green Man pub, the boozer where Dick Turpin is reputed to have drunk and where his self-appointed nemesis and chronicler, Richard Bayes, was landlord. Alfred Hitchcock used to come with his family when he was a kid, straight after Sunday mass. Daniel Defoe passed by. John Locke, the philosopher, heard of robberies in the forest while stopping off there 'and our Ladys hearts went pitapat'. Its name points back into an even deeper past, to the pagan figure of the man-tree, to the Green Knight of Arthurian legend, to Jack o' the Green, to some deep, dark, primordial mulch. It's now an O'Neills, selling reasonably priced pints of Guinness in a faux Irish setting and, frankly, who knows how much longer it will continue doing even that? It's a big ugly pile of a building and must be worth more as real estate than it could ever make pushing pints into the pink, shaking hands of ageing drinkers. And while its proximity to this roundabout probably doesn't do wonders for its value, let's be honest—*anything* in east London is worth more as flats than it is as a business.

There was a lull in passers-by, and I stopped taking pictures to stand still, trying to feel it with my feet. I knew I must look odd, but I had to make an effort to catch the vibrations. Nothing. Or at least nothing I could pick out amid the traffic noise around me, the sirens from the nearby hospital, the planes crawling overhead, a police helicopter bugging south across the sky from its base up behind Lippits Hill Lodge. But I knew it was there, knew how close I was to it. Not a ley line or anything like that. The four solid, traffic-clogged lanes of the A12 or Eastway or, as it was originally called, the M11 Link Road, running through the ground beneath me. When you're looking for the real routes of power, nothing is better than a road.

You can think of the M11 Link Road in two ways. Either it's the unintended consequence of the magical rites I've claimed were employed by Margaret Epstein to bring her son Jackie the petrolhead into the world—a cartoon phallus strapped onto the forest goddess. Or it's the unintended consequence of a series of badly made and subsequently botched planning decisions. Alternatively, you can choose to believe they're the same thing. The road was first mooted in the 1960s, part of an ambitious scheme to build four London orbitals or *ringways* cross-cut with a spider's web of motorways. The project—a science-fictional last hurrah for Big Government—fell apart without ever coming close to being realised. Gentrification was airdropping left-wing, middle-class fuss-makers straight into the poor neighbourhoods where no one was supposed to make a fuss. In 1970 the Homes Before Roads group polled 100,000 votes in London elections. In 1972 the Department of the Environment could issue a report stating 'the day of the supremacy of the motor car and the road builder has come to an end'. The following year the Greater London Council cancelled all work on Ringways 1 and 2. The government refocused its efforts on Ringway 3, out beyond the GLC's control, renaming its northern section the M16. It cut straight through the top of Epping Forest, affecting the village of Upshire, and soon the government found Parliament

Square occupied by the Upshire Preservation Society's tractors, their banners reading 'NOT EPPING LIKELY'.

One of the first cases of civil disobedience as resistance to motorway building came, in fact, at the beginning of the M16 inquiry in December 1974, just as Wally Hope and co. were packing up and leaving Stonehenge. John Tyme, an academic from Sheffield and leading light in something called the Conservation Society (of which, I suspect, he may also have been the only member), having had his points dismissed out of hand by the inquiry, started reading his submissions out loud all over again. And again. And again. 'I continued reading until the police arrived to escort me out. This was the first time the police had been summoned to a motorway inquiry. I was determined it should not be the last.' Eventually the M16 was rethought, connected up to the southern sections of Ringway 4 and renamed the M25. In order to avoid the problems which had stymied previous schemes, its planning approval was broken down into thirty-nine separate inquiries, each for a small, specific location, each highly technical, so that no grand narrative could be generated one way or the other. The thinking was small, piecemeal, technocratic, fundamentally boring, the sort of process that made men like John Tyme blahing on about 'consummate evil' look a bit silly.

The 1980s saw the pendulum swing the other way, a backlash led by Margaret Thatcher, who found in the motor car the perfect symbol of rugged individualism and atomised aspiration. In 1986 she opened the very last stretch of the M25, declaring that stretch of tarmac 'a showpiece of British engineering skills, planning, design and construction'. In 1989, ten years after coming to power, her government launched Roads for Prosperity, a six-billion-pound road-building project across five hundred locations including, of course, the resurrected M11 Link Road. The following year, Thatcher goaded the Royal Society, explaining that the glory of the garden of England required more roads for its upkeep and that Britain was 'a great car economy'. Her Transport Minister, Paul Channon, said he was 'the biggest road and bridge-builder since Julius Caesar', a wonderful example of the kind of kiss-the-ring, straight-up panto nastiness which made the Tory party of the 1980s and early 1990s so much fun to despise.

The Department of Transport, however, worked rather less efficiently than the Roman Empire at its height, and this was particularly true along the route of the M11 Link Road through Leyton, Leytonstone and Wanstead, where it cut through the southernmost stretches of Epping Forest. Rather than immediately bulldozing the houses as they bought them—either through compulsory purchase orders or just as people sold up, eager to get out—the buildings were left standing, presumably on the calculation that it would be cheaper and easier to do the whole lot at once. This presented a number of parallel problems. For one thing, the DoT wasn't meant to own housing stock. For another, squatters started to occupy some of the properties. A charity called Acme got in touch. They rented out cheap space to artists: would the Department consider letting them use some of the buildings temporarily? The approval was quickly granted. It can be argued, admittedly with a degree of hindsight, that if you want to flatten a load of old houses it's best not to leave them occupied by the most recalcitrant residents, wait until some squatters have moved in, then add artists to the mix. This combustible brew was completed when members of the Donga Tribe from the recently cleared Twyford Down road protests began to arrive.

All the same, for the situation to really catch fire the local community had to be enraged as well. Without that any explosion would burn out through lack of oxygen. The old common in Wanstead, George Green, down near the southern tip of Epping Forest, provided enough air for everyone. Local people had been under the impression—had been allowed to have the impression—that building the M11 Link would involve tunnelling under the Green. In fact, the plan was to dig out a whole chunk, build the road, then put a roof over the top with some grass on it. And to do this would mean pulling down a two-hundred-and-fifty-year-old sweet chestnut tree. When the green was fenced off in November 1993, the incoming roads protesters had the perfect focal point for local anger. A few days later, a large crowd of residents and protesters pushed over the fences, set up camp round the tree and built a tree house in its branches. The tree house had a letter box, a local girl sent them a letter and a local postman delivered it. A solicitor acting for the protesters went to court and argued that this meant the tree was a residence, requiring the DoT to serve an eviction notice to remove them. He won. The whole absurd, carnivalesque process took wing.

There aren't that many sweet chestnut trees in Epping Forest. You're more likely to find horse chestnut trees, which only produce conkers, good for playground bragging rights but pretty horrible to eat. There are a few sweet chestnuts in the GreenAcres Woodland Burial site up near Epping, where I visited Ken Campbell's grave on my way to Chipping Ongar. One sits opposite the wooden memorial representing his profiled face. Apparently there's a load down on Wanstead Flats, but I've never come across them. And the truth is, even if you can find them, sweet chestnut trees in England are almost always a disappointment, the nuts rather thin and pale and underdeveloped, much like our intellectuals. I would love to be able to tell you I found a great one to climb—that I sat high in its branches chewing on raw, chalky chestnut pulp, stuffing my pockets with shiny brown nuggets to take home and roast—but I can't. I was, though, becoming more and more caught up in the idea of tree-climbing, tree-perching, *tree-occupying*, as some sort of release, an escape. I kept thinking of out-of-body experiences, of ascending and staring down at my figure still sitting at the bottom, empty as a doll. It made no difference to the realities of any actual ascent, in which I was regularly reminded that no escape was possible. I remained earthbound, though with a heightened aspiration.

Looking at footage of the protesters up the tree—and round the tree—I found as much to horrify me as inspire. Long-haired dudes in porridge jumpers playing mandolins is not my idea of a good time. I can cope with all kinds of things, but not mandolins. By all accounts, though, everyone involved *was* having a good time, a million miles from the dry, procedural, technocratic borefest which had killed my political mojo at university. And the fun continued and grew, expanded and deepened, all the time as a rearguard action which could, to the uninitiated, look a lot like defeat. The eviction order on the tree house was granted in December 1993 and, with much wailing and recrimination, the protesters were removed and the tree pulled down. They regrouped at a house on Cambridge Park Road in Wanstead, also due for demolition and, on 9 January 1994, named it the Free Area of Wanstonia, in some kind of tribute to west London's Frestonia, a street full of squats due for redevelopment in the mid-1970s which declared itself independent from the United Kingdom. Heathcote Williams once again surfaces in our story. He had been Frestonia's Ambassador to Britain and now he declared, 'Long may Wanstonia reign and may it bring England back to civilisation.'

Wanstonia was cleared on 16 February 1994 and the action moved to Claremont Road in Leyton. This was the greatest stand of the anti-roads movement in relation to the M11 Link. Lasting almost a year, it became 'the longest and most expensive forced eviction in British history'. A whole street of houses had been abandoned, all except for number 32, the home of Dolly Watson, a ninety-two-year-old woman who had been born and lived her whole life there and who didn't want to leave. Some of the houses were used by artists under the Acme scheme, a few were already squatted, now the rest filled with roads protesters. The houses were knocked through and painted, the street itself was blocked off, planted, filled with barriers which were also sculptures, with sofas and stages and giant chessboards; a public space, a scraggy common. A café was started to generate funds and raise awareness among visitors. A miniature society formed, self-policing, 'an extraordinary festival of resistance'. As time went by, the houses were prepared for the inevitable final eviction. Concrete tubes were cemented through walls and roofs and into the ground so that people could *lock on*—handcuff themselves together or to embedded iron bars.

Dolly adopted the activists, defending them from media stereotyping and applauding their efforts. Eventually she collapsed during an attempted eviction and spent a couple of months in hospital, before being moved to sheltered accommodation, where she died shortly after. A scaffold tower almost one hundred feet tall and painted pink, named in her honour, was built out of the roof of one of the buildings (you have to see footage of the tower to believe how high and wild and delicate it looks. The sight of it gives me vertigo, a sick lurch in my stomach). Nets were suspended between the rooftops and surrounding trees. Barricaded rooms were built, tunnels and underground chambers excavated. Sound systems were brought in for free parties.

To the participants, Claremont Road represented a double-faceted resistance, first by slowing down the inevitable building of another road, and second, by demonstrating, almost as street theatre, another way to organise and live. Writing about Claremont Road, the architectural and urban historian Sandy McCreery characterises the theoretical underpinnings in terms of a dialectic between space and place: 'Place is lived space. If the space that capitalism produces is rational, ordered, mappable, controlling, anonymous, banal, and fragmented in its totality, then places are experiential, natural, transitory, confused, contested, unique, historical, and holistic.' The element of street theatre here was central: 'The No M11 campaign was a non-stop performance.' As well as building Claremont Road as some sort of stronghold, protesters went out daily to obstruct contractors working on other parts of the route, lying down in front of diggers, getting in the way, making a nuisance of themselves, even sabotaging plant and machinery by night, a practice called *pixying*. In addition, the media profile of the campaign was raised by stunts like climbing onto the roof of then Transport Secretary John MacGregor's house to unfurl a banner.

At the same time, a new front was opened up by the progress of the Criminal Justice Bill, which, alongside outlawing raves (any gathering of over one hundred people with amplified music characterised by *repetitive beats*, as the definition famously went) also criminalised a whole group of activities under the banner of *aggravated trespass*, a change of a piece with the Malicious Trespass Act of 1820 (and indeed the Legal Aid, Sentencing and Punishment of Offenders Act of 2012, which for the first time made the squatting of residential buildings illegal and hence a criminal offence). We will return to the issue of the criminalisation of trespass. In this context, the legislation opened up a much wider front, explicitly joining together roads protest, squatting and the free party scene which had burst into life in the late 1980s. Many people were involved in all three activities, but these were now explicitly grouped as one common enemy of the law and, hence, one common alliance. The government had helpfully joined the dots. In November 1994, the day after the Bill was passed in the House of Commons, protesters from Claremont Road succeeded in climbing up onto the roof of the Houses of Parliament to unfurl another banner and make their point to the world's media yet again. Later that same month, a huge army of private security and police moved into Claremont Road to clear it. It took them four days to get everyone out.

As the protesters had always known they would eventually be leaving Claremont Road, there was very little disappointment. They claim to this day that they made road building so costly and unpopular that most of the proposals in Roads for Prosperity were never carried out, although the counter-argument is that John Major's government had a very small majority and couldn't afford to alienate backbenchers with road-building through Nimbyshire. The No M11 group had in any case already morphed into a new form, one which focused more clearly on the new united front of ravers, eco-activists and squatters. They were called Reclaim the Streets. For their first event in May 1995, they crashed two cars together on Camden High Street, blocking the road, rolled in a sound system and partied for the rest of the day. Their most famous action involved blockading a section of the M41 and, under the skirts of stilt walkers, drilling up the surface and planting trees. Events both by the group and more generally inspired by the group have continued around the world until the present day. The clearest statement of their aims comes from Paul Morozzo, a No M11 veteran and key member of Reclaim the Streets. Echoing both Ubi Dwyer of the Windsor Free Festival and John Clare, Morozzo says: 'We are basically about taking back public space from the enclosed private arena. At its simplest it is an attack on cars as a principal agent of enclosure ... But we believe in this as a broader principle, taking back those things which have been enclosed within capitalist circulation and returning them to collective use as a commons.'

Anyone who has been involved in protest politics will know that you're not supposed to enjoy it. Part of your moral strength is drawn precisely from this lack of pleasure. As already mentioned, one of the innovations of the particular strand of dissent that developed at Claremont Road is that the people taking part appeared to be having fun. And we know they were having fun because they filmed it. They filmed everything. The actions at Claremont Road coincided with video cameras becoming cheap enough and portable enough to be available to more or less anyone. Maybe political protests became more self-regarding as well as more self-aware because of that. Whatever the case, while there's a certain amount you can read about Claremont Road, there's probably more you can watch. The place to start is with *Life in the Fast Lane – the No M11 Story*, a feature-length documentary put together by the people directly involved and now available for all and sundry to see, thanks to the power of YouTube.

Mick Roberts—also known as Old Mick—first appears about seven minutes into the film. Not as old as he seems at first, or seemed to his younger fellow protesters, Mick is bald with big sideburns, his hair slightly longish at the back. He's wearing a baggy, short-sleeved polo shirt with braces over it, the effect more shabby golfer manqué than skinhead. He and some earnest young men are shown barricading 15 Claremont Road in preparation for the coming evictions. There's an implication that it's Mick's house. Actually it's more than an implication. The banner that comes up with his name calls him a *Resident* and when he demonstrates a lock-on hole in the wall he says, 'This is what you have to do to try and stay in your house', before detailing to the others how he was offered fifteen hundred pounds to move out. 'They're spiritually bankrupt,' he goes on. 'They could offer me fifteen grand, thirty grand. They could offer me a hundred grand . . . They'd have to take off the roof first. They'd have to take people out the house first! 'Cos they're not going to do it. I'm not going to do it. Not any more. Not when you see your own area destroyed.' At that point he spots a slug on one of the wooden beams they've brought in. He picks it up carefully and carries it through the house to pop out of the back window—for luck.

Mick appears again and again throughout the actions that follow. He's there when the police form a line to hold back the protesters as contractors begin pulling down the chestnut tree on George Green: 'Your children'll see it . . . and they'll leave you,' he thunders. 'You'll be there in your old age with your *empty eyes!*' The crowd cheers. There he is again, sitting on the highest rooftop during the eviction of Cambridge Park Road/Wanstonia. That's him a little later, his hat now gone, hanging onto the ironwork protruding from one of the house's little mock spires, as the cherry picker which has him roped round the waist slowly pulls him up and away from the building, so that he looks like he's involved in making cheap special effects for a zero-gravity space movie. He's there on the roof of Transport Secretary John MacGregor's home and, later, sitting in a wood somewhere, reminiscing: 'We rather blocked their daylight out, apparently.' He seems different in these later clips. His hair has grown out, he's wearing a vest and a beanie, is tanned and relaxed—like he belongs. Then we're back with the action again: Mick lying in a net, rolling back and forth to check its strength, right up at roof height, the sky blue behind him. On 4 November 1994, Mick and three other protesters scaling the roof of the House of Commons, later, on the evening news footage, seen up on the ridge, bent over as he assembles a roll-up. Mick wandering round a room talking to himself as final preparations are made for the eviction of Claremont Road. Mick returning to one of the houses having smuggled food past the police. Mick up on the roof, doffing his cap for the camera. In that interview again: 'Each chapter, each page, has been written by young people who believe simply in what they're doing, the direction that they've got. It's not taking anybody out, it's not accusing anybody of anything. It's just quite frankly saying, We realise what the environment is and we're warning you that you're damaging it.'

It's hard to say exactly what it was about Old Mick that fascinated me. Perhaps I identified with his out-of-placeness. Mainly I liked the way he seemed to start off lonely—the sort of kind, quiet man who chatted with slugs—and ended up finding a home and a purpose there among the roads protesters. It seemed uplifting, I suppose; the ordinary guy whose life is changed forever and for the better by remarkable circumstances, the narrative from a film. I fantasised that he'd be living in a bender somewhere deep in Epping Forest. Then I found out he was dead, which both put the kibosh on my original plans and made me want to find out about him even more.

There is another video on YouTube, you see, which is also worth seeking out. It's from 2001 and it shows Old Mick's funeral. It's an astounding piece of footage. It starts with a panning shot across a massive gathering of the traveller/ eco-movement clans in a wood in Kent, all dreadlocks and flowers and shaven heads, little kids in baggy sweaters. The camera comes to rest on an open coffin. Mick is in there, still recognisable though a little green and twisted. The casket— rough, home-made—is picked up and carried shoulder high out of the trees, everyone falling in behind, dogs barking, the crowd yipping and yahooing. It is placed on top of a very large, very carefully built pile of wood. There's a short speech—slightly drunken, but slurred with emotion too— then they set the pile alight. Fireworks go off, someone rings a bell, everybody cheers. And the whole pyre burns, the flames strengthening, getting higher. You can hear people shouting: 'Go Mick, go / Go Mick, go!' and then, 'Love you, Mick!' And he's gone.

Throughout my researches Mick was always just gone. Whatever I came up with, he escaped me. Which is exactly why I wanted to hold on to him. First of all, I had to abandon my narrative of the gentle, lonely soul from

Leyton made heroic by the circumstances thrust upon him. It quickly became apparent that Mick was a slightly more complex character than that, a man of many facets, with an extensive criminal record and over ten years in prison. That was just the start of it. I began speaking to people who had known him and the stories proliferated. He was a cat burglar. He was the inspiration for the character of Dave, played by Terence Stamp in the Ken Loach movie *Poor Cow*. He'd blown up safes near Hollow Ponds in the 1960s. He had been connected to the Richardson gang and had ended up having to take the rap for something he didn't do—or at least something he wasn't directly responsible for. He'd escaped from prison multiple times. He'd been part of the breakout when John McVicar and nine other prisoners overpowered their guards on Dartmoor and ran away. He'd had his front teeth bashed out by prison officers trying to force-feed him when he was on hunger strike. He'd ended up in Durham on the high-security wing for repeat escapers, the one McVicar did his flit from. He'd faked madness to get transferred to Broadmoor. Or he'd been diagnosed as a creative psychopath. One or the other. Or both. And once he'd finished his time in the mid-1970s, he'd vowed never to go back or have anything else to do with mainstream society. He had lived in a removal van long before the birth of the New Age traveller movement and he had taken to squatting rather than claiming benefits. That was why he was in Claremont Road—not because he'd been Dolly's neighbour for thirty years, as I'd fondly imagined, but because he'd squatted one of the buildings after they'd all been emptied out.

Old Mick seemed to attract stories like some kind of narrative electromagnet. It was a heady, stinky mix of true crime, swinging sixties, prison break movies and eco-activism and I was thrilled by it. While a lot of people I'd written about weren't exactly canonical, they were hardly untapped either. But Mick, Mick was terra incognita! I found it hard to believe that his story remained untold. I thought I'd hit some sort of jackpot. A guy I'd worked with for years turned out to have lived at Claremont Road when he was fifteen or so and had squatted with Mick after it was all over, and we sat in a pub while he told me he had never known anyone else quite like him, a man who acknowledged no barriers or boundaries, who always did whatever he wanted, utterly without fear. He told me about going skip-diving with him at the back of supermarkets for the food they chucked away, and how Mick had driven it to protesters and squatters all over town and beyond in a van that couldn't get out of second gear and which had a sofa chucked in the front instead of seats. The look on his face was one of sheer relish, perfectly balanced at the point where admiration turns to incredulity.

I went down to south London and met Mick's younger sister, Nikki, a very youthful seventy and so clearly living on the bright side that she actually gave off a cloud of endorphins, and more than anything I wanted to please her and keep that hard-won cheer in place. I wondered later if Mick had something of this about him, this unassuming charisma, and looking back at the footage of young men buzzing round him, eager for approbation, I think he did. Nikki saw Mick's involvement with the roads protest movement as a final homecoming of sorts, the finding of a purpose, the conquering of his past. 'He was lost,' she told me. 'He was lost. Until he got focused on "Save the trees, save the world". It was a focus in life.' Seven brief years of it and then he was dead of cancer at only fifty-eight years of age, but seven years in which he became a kind of hero, an elder. 'He was a rebellious person,' Nikki went on, 'and he found his niche with this group of people. He really did. He lived how he wanted, he got the respect that he wanted.'

Nikki told me all about their childhood in west London, which was really the story of their dad, a man with an uncontrollable temper who, by fighting with the neighbours, got them evicted from their home. Later I found a letter that Mick had written where he retold this odyssey:

> When I was 14 years old we were evicted from our home in Hamersmith and had to live in a home for homeless families in Fulham, my older brother and Father had to live in lodgings as they were not allowed in the home with us, this left me with my Mother and my two younger sisters, this first break-up of the Family left a deap scar which took 2 years to heel before we were granted a maisonette in Roehampton, in this two years without a home or my Fathers guidance I ran wild and mixed with the wrong company & I got into trouble (house breaking) and was sentenced to an Aproved School for 3 years [sic].

Mick ended up in prison, Nikki said, because he took the rap for the Richardson gang, who he'd hung around with in Roehampton. He'd been made to hold a shotgun during a robbery and then told to take some jewels from a woman's hand. He was only eighteen and he got ten years. He had a terrible time inside, she said, 'he was good-looking and very vulnerable'. She thought he'd been raped, beaten up by the guards, he'd certainly got into fights, and he kept trying to escape. She remembered ringing her father from work to tell him Mick had gone on the lam again but not wanting her colleagues to know. 'The bird has flown the nest,' she said, several times, but her dad didn't know what she was going on about so eventually she had to tell him.

She remembered Mick as a boy climbing a tree in the park in Shepherd's Bush and getting stuck at the top so she had to get the park keeper. She remembered him telling her he was stopped in his car by the police with a big lump of cannabis resin and him saying it was a pencil eraser, showing them his art gear and saying, 'I'm a sketch artist.' She remembered his paper round, when he would dump most of the papers before stealing comics from the shop. She remembered visiting him at Homerton Hospital with her mother when he was dying and seeing him nicking a newspaper from Smiths, just the same after all those years.

She remembered going camping in Wales with Mick and him telling her about the stash of silver he'd buried there, his treasure. How he was always late for dinner, then would eat twice to make up for it. Taking her mum down to see him on an action in the Kingston area, where a row of poplars were to be cut down to improve the view from some luxury flats, how proud Mum was: 'people talking so highly of Mick'. Mostly she remembered how funny he could be, the mischief which could tip over into outright unreliability, so that even when he told them he had six weeks to live she advised her mum not to get too upset: 'You know what Mick's like. You can never believe what he says.' In among the stories of bad temper and thieving and unreliability, she kept circling back to her basic point: 'He had a magic about him, most definitely.'

Is magic just another word for madness? Was that the escape I was looking for and mortally scared of as well, a double-faceted flight into a new kind of prison? Early on in my investigations I told a friend, a counsellor, about Mick, about his lack of respect for boundaries, his failure to recognise even what a boundary was. 'Psychopath,' he said, and the conversation moved on, me left feeling slightly affronted by this response, but a little bit nervous too. What if all the stories I found funny, all the personality traits I was treating as admirable, were in fact the product of a pathology? Of course, pathologies are, to a large extent, socially defined, particularly social pathologies, and part of their role is to enforce and protect social norms. 'He was always a wild boy,' Nikki said. 'He would suddenly ask, "Oh, what's on the other side of that fence?" and he'd be away.' As a protector of the forest, though, an outsider defending that which is already *off limits*, these boundaries become so confused, so turned in on themselves, so everted and complicated, that to some extent they melt away, making Old Mick a pillar of his community, the man with the magic. Nikki showed me a short tract Mick wrote protesting about modern tree surgery's reduction of urban trees to 'executed stumps': 'in Victorian prints you see Avenues of light, from tall elegant trees, cascading cryptically about in paterns [sic] of rapturous chaos'.

I went up north and had lunch with Russell, a veteran of the Claremont Road protests. He met me off the train—a very tall, middle-aged man with a greying ponytail and a bad leg and one of those fleeces that's made out of recycled bottles. Russell was too clever not to be a little suspicious of my motivations and he greeted me with a certain reserve. But talking about the 'rapturous chaos' of life with Mick, his own enthusiasm for the man and the times they spent together got the better of him. Indeed, his apparently respectable front didn't hide his own pleasure in disruption for long.

It turned out that Acme weren't the only organisation renting buildings from the Department of Transport back then. Russell, a student at the time, had moved into a property managed by a housing association ten years or so before the protests. As well as the Acme buildings on the street, there was also a house rented out as a live-in studio space for ex-convicts with an interest in art. Mick had got very keen on sculpture when inside, and had fallen in with a whole crowd of prisoner-artists including the triple Koestler Prize-winner Robert Farquhar, aka Bob the Brush, who counted Sir Hugh Casson, President of the Royal Academy, among his patrons—as well as indulging an enduring penchant for robbing churches and cathedrals. Farquhar lived at Claremont Road for a while and would often stay, at least when he wasn't arguing with the other residents of the house. Eventually, Mick came and knocked on Russell's door—would he mind if he squatted the property next door to him? Russell was delighted. The neighbouring house was open and being used by local kids for parties and whatever else they couldn't do at home. Having someone living there would be easier.

Russell laughed a lot when he talked about Mick. He loved him dearly. Although he was loath to take too much credit, he played a key role in organising Mick's funeral and was still squatting with him right up until his death. He re-emphasised the importance of art in Mick's life, that he saw himself first and foremost as a sculptor. And if Jacob Epstein was an outsider in British society then so was Mick, with knobs on. 'He was always an outsider,' Russell said, almost as soon as we sat down. 'And he defined himself as an outsider. He wanted to live outside all the structures of the system.' His interest in art, Russell suggested, played a large part in kick-starting the artistic, playful, performance elements of the No M11 protests. It was Mick, after all (with Russell's engineering help) who built the famous hangman's gibbet on top of his house, the noose of which dangled out over the passing Central line train tracks. Not only was this the first piece of public art on Claremont Road, it gained almost mythical status among contractors and security workers, who believed that Mick would hang himself by it when the road was finally cleared. Mick and his fellow protesters did nothing to disabuse them.

It was Mick, too, who began the gradual blockading of Claremont Road—at the time used as a rat run and a police cut-through from Grove Green Road—by piling up tottering stacks of wood at the edge of the street and then gradually inching them towards the centre. He also pioneered Claremont's eventual attack on the boundary between *inside* and *outside*. Often, when local residents sold up or were asked to leave by their housing association, all their furniture would be abandoned in the house. Mick would move the living rooms, carpets and all, out to the front of the building and reassemble them. This became a theme of Claremont Road, the sofas in the street symbolising the collapse of the distinction between the domestic and the public.

Russell dates the inspiration for Reclaim the Streets to the complete blocking off of Claremont Road, which came after a huge bonfire melted the tarmac, enabling them to dig a trench in it overnight. A few people were sitting around in Mick's front room the following day as the idea for RTS formed:

'If we can close this road we can close any road.' He could hardly stop laughing about the time a representative of the Department of Transport had turned up and threatened that six named activists were to be sued for police costs totalling over seven million pounds. Mick put his hand up: 'Yeah, me—I'll take that one.' The idea was quietly dropped. As for Mick's commitment to the cause, he really did care about the environment and so on, but it was 'partly having a laugh. There was something about him that liked to poke at *any* authority. He didn't like officialdom in any of its forms.'

Mick could climb up anything, Russell said, a result of his days working as 'a cat burglar, basically, ex-jewel thief'. He had, he said, climbed out of Dartmoor Prison, up over the rooftops and down the wall, only to wander the moors for three days, completely lost, before blundering into the village where all the prison officers lived. He never claimed benefits nor paid taxes. He lived most of his adult life outside of prison in a variety of vehicles and squats, or stayed with girlfriends. He hated heroin and smoked cannabis 'like there was no tomorrow'. If someone needed something—a musical instrument, art supplies, food or brand new scaffold poles with which to build a one-hundred-foot tower, 'he would quite happily disappear and find it'.

While Nikki had talked about Mick's explosive temper, Russell only remembers him getting into a serious confrontation once—and that with the addict mother of a small child who got her daughter to prepare her fix for her. In fact, Russell says, Mick was very much responsible for keeping all their actions *fluffy*, in the parlance of the demonstrators. He was playful, much more interested in undercutting authority than confronting it. Apparently his model for the use of aggression was the rituals of Canada geese out on the Great Lakes of that country, who spend a lot of time puffing up and showing off their macho moves and very little time actually fighting.

Even the funniest stories illustrated Russell's point about Mick's liberatory tendencies, his failure to understand, let alone acknowledge, the basic concept of private property. He inherited a pet rat from another protester, for instance, and let it run loose until 'it made its own home'. Russell paused, chewing on his food while I imagined someone letting a rat *find its own home* in a series of scuzzy squats: 'It did get a bit much for him,' he eventually acknowledged. He would befriend dogs but never really wanted to own them, didn't seem to believe in ownership of anything, let alone any living thing. Whenever Russell's dog spent time with Mick it would come back without its collar on. If Nikki suggested Mick as naughty schoolboy or *Beano* character, Russell saw a natural anarchist. When serving a long stretch in prison 'he developed lots of knowledge about getting about on the sea', Russell told me. 'He had this dream that when he came out he'd just get on a boat with a compass and a sextant and go that far outside of *everything*.'

I love a breakout. I always have. My two favourite Christmas movies when I was a kid were *The Great Escape* and *Hannibal Brooks*, in which Oliver Reed saved a zoo elephant from the Nazis by leading it across the Alps to Switzerland. I was obsessed by flight, by birds and biplanes. Stories about Houdini fascinated me, and when they made the old radio show *Sexton Blake* into Sunday-night TV entertainment, what stuck with me was the moment at the end of each episode when the detective asked his assistant to tie him to a chair so he could get free. Seen within this context, the appeal of Mick Roberts is obvious, as is the fascination of the forest, this *outside* zone, this fugitive land and hiding place.

Then again, why am I so obsessed with escape? I had a happy childhood, so far as I can tell. I wasn't kept locked in a coal cellar or interfered with by my rugby coach. And actually, when you take the time to think about it, you *can't* go 'outside of everything', because in that case, what are you *in*? There was ambiguity, contradiction even, at the root of my position, at the root of my whole project, in fact, and calling it enantiodromic didn't change that. The escapes listed above could broadly be described as heroic, but my own were largely pusillanimous in motivation. I have a highly developed flight mechanism—also known as physical cowardice—which has presumably steered generations of Ashons away from danger and allowed them to procreate, cowering in a ditch, while braver souls disembowel each other. And hiding in a ditch is still a form of escape, though not one met with too many fanfares. What appealed to me about Old Mick wasn't so much that he was braver than me but that he was completely beyond notions of courage or cowardice, essentially *unbound*—a transcendent affirmation of my forest philosophy. He wouldn't be averse to hiding in a ditch if the situation demanded it, but if he did, you can bet that, rather than soiling his breeches, he'd find it hilarious.

I went to visit the amputated remains of Claremont Road, now a short cul-de-sac ending abruptly in a two-tone wall, sand brown and brick red, beyond which could be heard only cars. The wall was so blank, so incongruous, it looked as if it should be the set for some kind of stunt or trick, a feat of evasion, of climbing over or digging under, of disappearance or border-crossing. Instead of finding lift-off, though, all I could think of was the twenty-foot drop into seething traffic beyond. Old Mick hacked out a laugh and was gone.

CHAPTER 8

Theydon Bois | Ambresbury Banks

I wasn't looking for Theydon Bois Golf Club, exactly, but I found it anyway. I burst through a hedge, dazzled by a low winter sun, and found myself out on short-cropped, viri-descent grass, surrounded by a semicircle of ageing sports casualties with tasselled shoes, pastel sweaters, baseball caps and five irons or drivers or putters or whatever—harbingers of mortality, anyway: mine through battery, theirs through boredom. I didn't mention my un-vestigation into the death almost forty years before of a hippy drug dealer. I didn't think it would help. Instead I enquired after the main road and, if they weren't exactly friendly, they were happy to point me in the right direction. There was something awkward going on all that morning, Theydon Bois itself a viper's nest of low-rise respectability which reminded me so violently of the village I grew up in that only the promise of a very milky coffee from the gussied-up patisserie stopped me from breaking into a run. It was all internal, but the fear was strong.

I was searching for Wally Jeff Moonlight aka Wally Moon. At some point, Penny Rimbaud had used the example of the death of a follower of Wally Hope's in Epping Forest as an illustration of the intimidatory atmosphere around after Wally's death, as well as to suggest possible drugs trade links between organised crime in Brighton and the Essex CID. Mr Moonlight's passing had been unusual enough, it turned out, to be reported in *International Times*, though Penny's chronology was screwed and it happened before Wally's incarceration: 'He was found in Epping Forest on his back with his left hand tied to a tree, a joint (unlit) in his right hand, and a bottle of wine (half full) next to his elbow.' The atmosphere of the era can be gleaned from the *West Essex Gazette* of 13 December 1974, which reported Chief Inspector Robert Storey telling the inquest that Geoffrey Moonlight may have decided to take his own life: 'He lived well into the drug scene and it might have been the way he wanted to die.' The claim was disputed by a friend, and by Geoff's father, Charles Moonlight, who said he had been 'extremely cheerful' when he had left their house. An earlier report in the same newspaper stated that the body was found 'on a footpath near Theydon Bois Golf Club', and for some reason I thought this would be enough information for me to find the spot. In case this hasn't yet become fully apparent, I reiterate: I like the idea of a pointless jaunt.

TB, as I've come to call it, is the next village up from Loughton, smaller and just as close to the forest but—lacking the size for a bohemian enclave—somehow even more monocultural, if such a thing is possible. The place put me on edge. I got off the road as soon as I could and followed a footpath round the border of an estate, along the back fences of the gardens, a muddy ditch of a walkway hemmed tight between two sets of property, houses to the left, farmer's field to the right. It was one of those days when whatever I did made me feel like a lurker, an undesirable, forever in danger of being told off or chased away or reported to the police. When the path finally peeled off from the houses there was temporary relief in a broad green hill starred with decrepit oaks, before I was forced round the perimeter of a large farm. Then the golfers, then the track alongside the golf course with no sign of any footpath, then across the main road and into the forest proper, only to find more golf course on the other side, a flash of baize beyond the brambles—the kind of snarl round my feet which seemed the perfect resting place for a corpse.

I t's an attractive idea, Penny's *just happening*—that sudden, post-op revelation he told me about, his instantaneous, Zen-soaked serenity. It's even more attractive if you disregard the seventy years of struggle and striving that came before it, if you conveniently overlook the fact that you have *not* to be looking for it in order to find it. If you shrug off the uncomfortable truth that you're not even sure what *it* is. And yet, there I was, looking for something to just happen, while trying to look like I wasn't. Trying to sidle up on it by chance. Hoping to stumble across some sacred grove out in the forest where all my wishes would be made clear to me so that they could then be granted. *It just happened*—the easy magic of the lazy non-believer. Of course it appealed to me. As long as I could keep up the hardcore disregard that would be my kind of spell. After my second visit to Dial House, instead of trying to meditate I took up running again, a slow, stomping, awkward presence panting round the local streets, squelching beneath trees. I did it more and more, a compulsion. And I knew full well I wasn't running towards anything. I was running away.

I don't want to overstate any of this. I hadn't been diagnosed with cancer. I didn't have a major drug or alcohol problem and neither did my children. My wife hadn't left me and I hadn't left her. I wasn't mourning the death of a parent, addicted to slot machines, driving myself to the brink with internet porn or crushed by a monster of debt. To the best of my knowledge, I didn't even have bad breath. If you imagine this type of book pictured as one of those old graphic representations of world mountain size, then I hadn't even made it up Snowdon, let alone got near to an Everest of upheaval. But it was still cold, the wind whipped in from the north and the cagoule of my inner resources was hopelessly inadequate, something left over from the late 1970s, unbreathing, both holding in my sweat and letting through the rain. People die on Snowdon every year. They also get blisters, twisted ankles and, on occasion, sunburn. I was below the treeline, though, so at least I had shade.

I picked up a path through the scrub, the sun obstructed by trees, the light dark and bright all at once and close together, so that it spotlit the old lady on the track in front of me—the fine white hair and pale skin, the shape of her skull, the knuckles of her hands like lumps of rock—as she stopped, turned back and looked at me. Her Alsatian stock-still, looking too, its eyes round and very black as it sized me up. It waited a moment, allowing time for my imagination to fill the gaps, then began slinking towards me, keeping itself low so that its shoulder blades stood up above its spine, quiet, purposeful. I was being hunted.

I've never liked Alsatians. Back when I was growing up they were the cultural equivalent of pit bulls, with the added baggage of their role with the police. They used them at Orgreave to bite striking miners. My bags were sniffed by them on disembarking from the Amsterdam ferry. I filed past them on more than one demonstration as they barked, up on their back legs, straining at the leash. My theory is that as the image of the police became more politicised so too did the image of their dogs, so that a disenfranchised working class turned to breeds which showed more clearly their defiance of authority. Either that or they just decided pit bulls were scarier, better at biting and fighting and hanging from trees, more redolent of the US projects and a real, armed, saggy-pants underclass. Terrified of them all, I'm not the man to ask.

I stopped walking and waited for the old lady to call the animal back. She didn't. We stood, two fixed points, the dog a vector between us. I tried not to look at it while keeping close tabs on its progress, holding only shoddily to the idea, often imparted to my children, that canines are hungry for attention rather than just hungry. When it was about four paces from me it stopped again, raising its head, and stared at me, completely unreadable, blocking the path. None of us moved until, in a slightly whiny voice, I asked her if she could call it back. She didn't appear to do anything much. Perhaps she turned away and started walking again. Perhaps the dog got bored of me. For whatever reason, it swung round and began moving off.

I had a problem. There was only one path and the woman and her dog were on it. I didn't much like the dog and nor did I want her to think I was following her through the woods. The guilty feeling which had been stalking me since I got off the tube in Theydon Bois took a step forward and climbed on board. I looked like a weirdo, a rapist, at best a flasher. What other reason did I have for tracking an old lady through the undergrowth? No wonder she'd let her beast come and get a fix on me. I was lucky she hadn't ordered it to go for my throat. Rattled, I assessed my options. I could turn round and go back, but somehow that would be even more suspicious. I could stand here for another ten minutes, caught between a housing estate and a golf course, pretending to have seen something really interesting. Or I could try to get past and out into the open woods.

Luckily, however it may seem, there's never only one path through Epping Forest. After blundering off at a forty-five-degree angle to the main trail, my feet lifted dressage-high above the scrub, I soon picked up another track, wiggly and less well defined but good enough that I could accelerate and try to navigate past her. That was my plan—stride out and skirt round in a broad crescent, pick up the path beyond her and move quickly up to where it intersected with one of the forest's big bridleways. It had the whiff of wishfulness about it, of hopeless hope more than expectation, a lack of conviction which left me exactly where I deserved to be. I came out ahead of her but near enough for it to seem creepy. She was moving quicker than I'd allowed for, quicker, certainly, than when she'd stood staring at me. She may have thought I was trying to cut her off at the pass.

In an enantiodromic turnaround, I was no longer clear whether I was supposed to be the hunter or the hunted. I picked up my pace, trying to leave her behind, all thoughts of poor young Geoff Moonlight lost in this pell-mell retreat, huffing along, hot now, ruffled and discomposed. Reaching the bridleway I turned a sharp left, tracking back at an acute angle to the path I'd left so that not only could I see the old lady and the dog near her waist, but see them cutting off the corner, inexorable, gaining on me, like the posse in an outlaw's most fevered sleep. I had allowed myself to think I had lost them and in that moment I knew I needed to piss, regretting my large milky coffee and its diuretic imperative. Now I tramped fast along the bridleway, being overtaken by nice ladies on horses, greeted by nice lady walkers moving in the other direction and, still, pursued by my nemesis and her hound. As soon as I saw what looked like a path I veered off right, searching for somewhere discreet to empty myself. The area is called Epping Thicks for a reason. Holly bushes have taken a decisive hold up here, so it's not as easy to choose your own route as it is further south. I worked my way in, ducking and twisting, and even before I started to pee I knew I'd need to reverse. I was trapped. There was no way through.

Back to the path, back to the old lady, back to the Alsatian—nearer, faster—then off the path again and back again, two times, three times, each one more desperate, each one more frantic, before—at last!—the holly started to thin and I found a trail through and away from her for good. In an aimless, relieved zigzag, I drifted to the northern wall of Ambresbury Banks, the dread and self-spooking finally dissipating. I was over the far side of the hill fort, as far as possible from the path that touches it, so there was no one else about. The earthworks rise higher here than at Loughton Camp, a gigantic ripple of soil, as if a god has dropped a slow-motion rock into the geologic pond. I sat on a fallen tree and drank some water, ate a banana, placing the skin inside the coffee cup I was carrying from earlier. Local tradition had Ambresbury pegged as the site of the last stand of Boudicca when her revolt against the Romans was finally defeated, but apparently Victorian archaeologists definitively disproved this—which is somehow typical of the Victorians. Maybe it was my feeling of relief, or the low sun slanting through the trees. I'm really not sure. But it was as I sat there that it came to me. What kept drawing me back to the forest was the fact that I didn't have to stick to the path. And the reason I didn't have to stick to the path was because it wasn't a thin right of way through private property. I wasn't scared of getting lost. I was scared of trespassing.

In this if in nothing else, I was walking in the footsteps of a poet. John Clare's fear of trespass haunts his work: the fences, the signs, but also the anxiety of straying, of being caught by accident without knowing what it is you've done, even the anxiety of being *thought* to have strayed. Just as significantly, though, there is a part of Clare which *wants* to transgress, to cross that boundary and, thinking about it now, it's hard not to see something charged, something erotic in these games he plays, both in relation to Mary Joyce and to the land around him. He is forever stepping over fences where he shouldn't step, furtive, eager, getting off on not being caught. Getting off, even more, on the chance of being caught. It reminds me of Philip Roth's definition of the writer as a *closet exhibitionist*. Seen in this way, Enclosure represented an opportunity for Clare as a writer even as it made a mess of Clare as a peasant, his bivalent soul finding both expression and repression in the privatisation of land. It wasn't just his subject, it was what animated him—the forbidden, caressing the untouchable. It was him, in point of fact, who made Mary Joyce forbidden, hence in some way eroticising her in perpetuity, so that when he marched home from Epping it couldn't have been to claim her but only to come close enough to touch what he wasn't meant to touch, some kind of voyeuristic sublime. His diffidence is less a result of fear or some failure of will, but of the contradiction which lies at the very heart of what he is. There is an immense bravery—by necessity unacknowledged—in exploring the borders of this forbidden territory, in sneaking over to report from the other side, in relying always on Berger's 'ambiguous answer' (the 'truth as an uncertainty' of the peasant world view), denied him only by the full stop of his muse's death.

By comparison, my own problem with trespassing is rather insipid and I can only explain it by telling you a story. The village I grew up in was just to the south of Leicester, a small city right in the middle of England recently made famous by the discovery of the bones of Richard III, the last Plantagenet king, beneath a car park, and then made more famous—though perhaps only briefly—by its overperforming football team. Countesthorpe itself was a crooked H of old houses following the main roads, filled out by estates built in the 1960s, a kind of aspirational toytown graft, all pale brick, white wood panelling and large windows. The village was named for a niece of William the Conqueror's after he gave her some land here as part of her dowry in the ongoing dodgy deal which was the Norman invasion. Around seven thousand people live there now and, as only one new estate seems to have been built since my childhood, I don't think it would have been much less than five thousand back in the 1970s. Overseeing law and order for the inhabitants was the village bobby, PC Connor.

He already seemed like a remnant, even then; an angry, dyspeptic version of Dixon of Dock Green, raging against the slow erosion of his authority, old and pissed off about it, his hair a severe grey short back and sides, his buttons polished, his trousers a cut from the 1950s, dismounting from his black sit-up-and-beg bicycle with the cautious deliberation of a naked man climbing off the back of a leopard. I asked a friend for his reminiscences of the man and all he could think of was how much I hated him. I don't remember how old I was when this antipathy began, but I was still riding my first bike, the one I had before I moved up onto my sister's old red and white thing, which became my Sopwith Camel and on which I flew many missions against Baron von Richthofen. I'm guessing I must have been seven, maybe a little younger.

The ice cream van didn't come down our road but instead used to park up on Aspen Drive, from where we could hear its bell summoning us. Our road was old and made to seem older because it was private and hence hadn't been resurfaced since some time in the 1930s. Aspen Drive joined on to its far end, a few bollards separating us, with Steven's house on the right, a willow filling the front lawn, and Barbara's on the left, where I went after school when my mum was at work. And outside Barbara's, the ice cream van. On this particular occasion my mum was on the phone and I had to wait for her to finish before I could ask her for some money. By the time she'd dug out the change and I'd dragged my bike outside and begun to twirl my way down the potholed, bumpy road—jagged black stones bigger than old two pence pieces poking up from the tar—the van was already moving off.

At that point I should've returned home. The end of Aspen Drive was as far as I was allowed to go. But having already got the money, having had to wait, having felt the anticipation of the creamy, chemical vanilla taste of Mr Whippy—my mouth watering even now, thinking back to it—I carried on.

Somehow I knew the van's next stop would be in Springfield Drive, two streets over, and I knew if I pushed it I could get there in time. It can't have been very far, a few hundred metres at most, but it seems a long way in recollection, my little legs spinning out, the bike inching along, its route a wriggle of over-swerves as I strained for speed. The chimes stopped again, so I knew the ice cream man was already serving a diminishing queue of kids. When I reached the corner of the road, sure now that he would be gone again at any moment, I cut across the verge. That's how I remember it happening, anyway, though it doesn't make much sense. The verge was a strip of grass outside the corner house's side fence, but for it to count as a shortcut I would've had to carry on and plough right across the front lawn, escaping between the rose bushes on the far side, and I cannot, in my wildest imaginings, see my childhood self doing that. Maybe I lost control and went across it by accident. Maybe I felt safe further from the road. Maybe I didn't think through where my route would have to take me next.

It didn't matter. My progress was halted by PC Connor emerging from his house, the corner house, and standing on the path to block my way. I remember him towering over me in his police trousers, police shirt and police braces, his helmet and jacket inside (though I may have invented the braces, and I seem to remember or have also added that he was having his tea). PC Connor barked at me to stop. He told me that I was trespassing on private property, his property. He told me that I had broken the law and that if he felt like it he could arrest me and take me to the police station and my mum and dad would have to come and get me. He told me I was never to ride my bike on his property again, that he was watching me. He made sure I knew I was a criminal. I can't remember if he waited until I was in tears or left before they began. Either way, by the time he was done with me the ice cream van was gone and I was sobbing, slumped over my handlebars. I never told my parents. I was too scared. I remember lying awake worrying about the consequences of my crime.

He was wrong, as it happens, or lying. He couldn't arrest me for trespassing as in most cases trespass is a civil offence under common law. The term itself is the legal French equivalent of the Latin *transgressio*—wrongdoing (which takes us back to Clare again). It seems to have been another Norman import, in this case from Roman law, a system whereby an injured party could sue for damages to replace the Anglo-Saxon use of Compromise, a negotiated settlement. In its original form it referred to any injury done to a person, chattels (from the Old French for cattle, meaning movable property) or land. At this time, however, there was no clear distinction between a punishable crime and a civil offence: 'Our earliest examples seem all to be cases of undoubted violence with a strong criminal element. The plaintiff has been beaten, wounded, chained, imprisoned, starved, carried away to a foreign country, and has suffered many "enormities".' Just a normal Saturday night in any English town, then. The same is true of the narrower—and for us, more relevant—class of cases covered under 'breach of close', breaking a man's close or fences, aka *quare clausum fregit*. At first concerned with 'violent invasions by marauders, accompanied in most cases by serious assaults on the owner and his servants', by 1466 and the famous Case of Thorns, it came to cover liability for the unintentional consequences of your actions. The defendant had cut thorns from his hedge, they had fallen on the other side and in recovering them he had damaged his neighbour's crops. His defence that he hadn't known they would fall on his neighbour's land was rejected, and not just because it was stupid.

Trespass was *not*, though, a criminal act. You might think that if you were looking to protect your property and had control over legislation (as the landowners clearly did until well into the twentieth century) then your best bet would be to make it illegal. In fact, the civil status of trespass conveyed some advantage to landowners. Aimed at redress for the wronged rather than punishment of the wrongdoer, the weight of evidence in a civil case only has to be, on balance, *in favour* of the plaintiff, rather than being beyond reasonable doubt, a far more demanding standard. If the defendant should turn out to be poor and hence unable to pay any redress so granted, the punitive measures taken against him or her would quickly ratchet up. In addition, whatever actions the trespasser might take in the act of trespassing, or as a reason for trespassing, were criminal. Hence while trespass itself may have been a civil offence, if you picked up some wood or allowed your cow to graze on fenced-in grass, you were stealing. For a civil offence you could be made to pay damages; for a criminal offence you could be convicted and transported. Any *use* of enclosed land was illegal. If you had previously relied on that land for your subsistence, then you had a problem. The story of Enclosure is the story of the criminalisation of most of our rural population. Enclosure is, in effect, an exclusion and this exclusion was enforced by a full range of punitive measures: damages, incarceration, transportation, even death.

Between 1700 and 1870 it's estimated that fourteen million acres (twenty-two thousand square miles, almost half the country) were enclosed in England, most by Acts passed in a Parliament full of landowners. As the agricultural writer and campaigner Arthur Young put it in his 1801 report, 'by nineteen out of twenty Inclosure Bills the poor are injured, and some grossly injured'. During the reign of Henry VIII, land was being grabbed from common usage for raising sheep, causing the gradual collapse of the common field system in which peasants tended their own strips of crops. By the 1700s, this whole system of farming was under attack as being uneconomic and irrational: 'There were many whose eyes glistened as they thought of the prosperity they were to bring to English agriculture, applying to a wider domain the lessons that were to be learnt from the processes of scientific farming.'

But beyond common farming practices, the heyday of Enclosure saw a concerted attack on the commons and wastelands on which tenants gathered fuel, trapped rabbits, collected mushrooms and berries, fed their cows, goats, geese and so on. Not only was it held that this land was inefficiently used, it was also asserted that its existence was morally harmful to those who relied on it. In a discourse echoed today in newspaper narratives about *benefit dependency*, the ruling classes complained of the fecklessness and dishonesty which was bred by access to common land. In 1720, the Duke of Montagu stated that peasants used wood-gathering as 'Cloaks for ye Greatest Villainies, in destroying the Wood and the Game'. Billingsley, in his *Report of Somerset*, claimed that 'in sauntering after his cattle', the commoner 'acquires the habit of indolence'. Charles Vancouver, in a similar report for Hampshire, opted for describing the commons as 'nests of sloth, idleness and misery'. Arthur Young, in his role as an editor for the Agricultural Board, advertised the view that Epping

Forest itself was the 'nursery and resort of the most idle and profligate of men'. For their own good, these people should be denied access to the kind of *free lunch* that prevented them from wanting or needing to work a full day for a paltry wage.

Enclosure was the defining crisis of Britain's working poor. Despite the supposed increase in efficiency resulting from Enclosure, food prices went up. Meanwhile the wages of farm labourers went down. Factor in the loss in common benefit (one study estimates that collected firewood might have been worth as much as 10 per cent of a farm labourer's wages, and a cow providing a gallon of milk per day during season might be worth the equivalent of half a labourer's wages) and suddenly it became nigh on impossible to live in the countryside, certainly without resort to theft. Enclosure and the Industrial Revolution were two mutually supportive facets of the same phenomenon. The proportion of the population living in rural areas dropped from 65 per cent in 1801 to 23 per cent by 1901. These people poured into the cities, looking for work in the new factories, living one on top of the other, filling the poorhouses, sleeping rough on the streets, doing whatever they could to survive, so that the American author Jack London, visiting the East End in 1902, could write, 'if this is the best that civilization can do for the human, then give us howling and naked savagery. Far better to be a people of the wilderness and desert, of the cave and the squatting-place, than to be people of the machine and the Abyss.'

Successive governments introduced increasingly draconian laws to cope with the crime wave created by their own policies, a crime wave driven, first and foremost, by starvation. To fully understand John Clare's fear of being mistaken for a poacher, it's worth looking at the successive Game Acts introduced in the first twenty years of the nineteenth century. In 1800 the punishment for poaching was a maximum of one month's hard labour in a House of Correction. By 1816 it was seven years' transportation. Between 1827 and 1830 it is estimated that one in seven of all criminal convictions were under the Game Act. In 1820, trespass was, briefly, criminalised, or as good as. The Malicious Trespass Act stated that any person convicted before a single magistrate of doing wilful damage to building, hedge, fence, tree, wood or suchlike was to pay damages of up to five pounds and if unable to pay be sent for hard labour in gaol or a House of Correction for three months. Males under sixteen would automatically be sent to a House of Correction for six weeks' hard labour. In the following seven years, 1,800 boys in Warwickshire alone were committed under this law. And, as the magistrate was often the local landowner, 'malicious damage' could be something as small as breaking a stick off a tree while walking down a road. Wandering through the woods, then, was a dangerous undertaking. A magistrate of the time, Sir William Dyott, summed up the impetus behind these laws and the harshness of their application when he said that 'nothing but the terror of human suffering can avail to prevent crime'.

All of this nineteenth-century legislation, though, was merely an intensification of the theme established by the Acts that were passed back at the start of the Enclosure era in the early 1700s. George I was the grandson of Elizabeth Stuart and Frederick, Elector Palatine and sometime Winter King of Bohemia. Although over fiftieth in line to the throne on the death of his second cousin, Queen Anne, he was her closest Protestant relative and, under the terms of the Act of Settlement introduced by William of Orange in 1701, no Catholic could be allowed to inherit the English throne or marry the monarch—a law that stayed on the statute books until 2013. George was crowned in 1714 and the Whigs used the opportunity to consolidate the power they'd been building since they supported William's Glorious Revolution in 1688. Over the next nine years, five major Bills were passed that were to limit definitively the possibility of resistance to Enclosure. The Riot Act of 1715 meant that local authorities could declare any gathering of twelve or more people unlawful. The Transportation Act of 1717 established the transport of convicted felons to America as a new standard of punishment. The Tailors' Combination Act of 1721 was the first of a series of combination acts which limited workers' rights to unite and negotiate collectively. The Poor Relief Act of 1722 enshrined the idea that the workhouse should function as a deterrent, that relief for the poor should be on some level punitive. Last came the Black Act of 1723, which made it a crime punishable by death to hunt deer, poach hares, coneys or fish at night while armed and disguised, but also to send anonymous letters, cut down trees, shoot at a person, demand venison or money or to help anyone known to have committed such offences. The breadth and severity of this last Bill has been summed up by the criminologist and historian Sir Leon Radzinowicz: 'There is hardly a criminal act which did not come within the provisions of the Black Act...and all were punishable by death. Thus the Act constituted in itself a complete and extremely severe criminal code.'

When I left Ambresbury Banks I crossed over the main road, scuttling between the cars which thunder along this arrow-straight section way over the speed limit. The biggest path through the woods up here runs past the Banks, so this side of the road, known as St Thomas's Quarters and lying close to open farmland beyond, is among the quietest and least visited. I was here to hunt deer. There are two types in the forest these days. The small, stunted muntjac—that looks more like a wild pig and sounds, as it tip-taps along in front of you, like a demonic child with razor-sharp teeth tracking you through the undergrowth—is mainly found slightly further south, nearer to Chingford. It is solitary and bad-tempered and has a reputation for going through bins, so suburbia suits it. The fallow deer, on the other hand, looks and acts like an actual deer—white tail, nervousness, antlers, pronking—and, in my experience, is found most often around here and in the fields beyond, though there is a sanctuary back over towards TB, where my journey started, if you're more interested in being certain than being surprised.

I'm sure if you know what you're doing you can actually track and/or stalk fallow deer. You presumably look for footprints and chewed shoots and sniff the air for their musk. My method is to wander around very quietly for a while in the hope I bump into them. Sometimes it works, sometimes it doesn't. It's the anticipation that makes it. On one occasion a whole ribbon of them walked across the track in front of me, of every size, and when they saw me stopped and staring, they stopped and stared back and none of us could work out who was more afraid. Today, though, this crude magic deserted me and I lumbered about trying to be quiet and unobtrusive in terrain apparently designed to make me noisy and obvious, dead leaves and sticks crunching under my feet, branches catching and scratching at me and pinging back, jays and blackbirds rasping and chittering tidings of my presence, squirrels bouncing back to the safety of trunks, wood pigeons scudding out of the treetops in panicked clouds.

The Black Act was introduced in a rush and amid considerable panic—an early example of a media-driven crisis—to counter the supposed threat of a gang known as the Waltham Blacks. Unfortunately for my purposes, they weren't from Waltham Forest, but Waltham Chase down in Hampshire. In October 1721 sixteen of them broke into the Farnham park of the Bishop of Winchester, shot one of his gamekeepers, carried off three deer and left another two dead. The Bishop, 'naturally of a warm and cholerick Temper', had four men arrested, two of whom were convicted. In response, the Blacks returned to the Farnham estate in even greater numbers and carried away another eleven deer, leaving the same amount dead and parading their haul through the local town at seven on a market day morning.

The Blacks—so-called because they blacked their faces as night-time disguise—continued their raids until the park was almost emptied of deer, then turned their attention to the bishop's land at Waltham Chase. From here they branched out, attacking other landowners and sending letters threatening to burn out anyone who stood in their way. A Mr Wingfield charged some peasants for carrying away more wood than he had allowed for from his land. The Blacks came and stripped the bark from his trees, maimed others and left him a note, 'to inform the Gentleman, that this was their first Visit; and if he did not return the Money receiv'd for Damage, he must expect a second from King John of the Blacks'. During this time, King John even made a public appearance to protest his loyalty to King George at a pub near the Chase, 'but 15 of his Sooty Tribe appear'd, some in Coats made of Deer-Skins, others with Fur Caps &c, all well armed and mounted: There were likewise at least 300 People assembled to see the Black Chief and his Sham Negroes'.

As can be gathered from this recollection, there was more to face-blacking than mere disguise. You could read into it a theatrical display of savagery but also, perhaps more fancifully, some kind of identification with the plight of African slaves, at that time being shipped to the Americas in their hundreds of thousands. No one would inform on this group despite substantial rewards and immunity from prosecution being offered. Dragoons were summoned to patrol the local area and the situation became so heated that King John appears to have disbanded the group. Most of the attacks up until this point had been on prominent Whigs. E. P. Thompson argues this was a direct result of Whig landowners using George I's lack of interest in hunting—and hence in the forest—to advance their own interests: 'The Hanoverian accession had withdrawn the actual presence of the monarch from the Forest, thus enhancing the influence of those noblemen and officials who derived their authority from "the Crown".' Under Queen Anne, forest regulation had been relaxed; now these landowners—who also happened to control the government—were encroaching ever more on commoners' rights. As King John told the *London Journal*, the Blacks 'had no other design than to do justice, and to see that the rich did not insult or oppress the poor; that they were determined not to leave a deer on the Chase, being well assured it was originally designed to feed cattle, and not fatten deer for the clergy, &c'.

Whether for genuine reasons or for political expedience,

Walpole and the Whigs saw the hand of Jacobitism in the Blacks' actions and pronouncements, as if their aim was to return a Stuart to the throne and the best way to achieve this was by poaching deer. It was this essayed connection which justified the introduction of the Black Act to crack down on their potentially treasonous behaviour (though there were already plenty of laws for treason). And although none of the original group was apprehended, a second version of the organisation, something of a tribute band, was put together and this gang upped the stakes still further by raiding the king's own deer from Windsor Forest.

The leadership of this group were, however, betrayed by some of their own members and then trapped when tempted by government agents to come to London to give evidence against one of their old enemies. In the end, thirty-two members of the Blacks were arrested, six sentenced to be hanged for the murder of a gamekeeper, another six transported. Seven more, probably from the original gang, were captured after a further raid on Farnham and of these, three were found guilty of Wilful Murder and the other four under the auspices of the newly introduced Black Act. All of them were hanged. And that was the end of the Waltham Blacks.

Animals make tracks, humans make paths. But paths in forests model the complexity of their surroundings. They *branch*, then they branch again. The artist Olafur Eliasson—best known in the UK for *The Weather Project*, his gigantic sun in the Turbine Hall at Tate Modern—tried to capture some of these compacted meanings in his 1998 installation *The Forked Forest Path*, which turns the art gallery into a wood, branches rising up round the path to make a tunnel through which the viewer walks. It isn't one of his strongest works, though I've only seen pictures of it and maybe it's different to stand in it and feel surrounded; maybe it catches a whiff of the panic of feeling lost. But the clean lines which made *The Weather Project* so striking seem to me to work against the messy physicality of the path—something made on the ground, not in the air, something which, harking back to the word's Germanic roots, is *trod*. In that sense, Richard Long's famous and self-explanatory *A Line Made by Walking* comes closer to expressing this *pathness*, though in its monomania, its ramrod-straight intensity and the fact that it goes nowhere, it is in other senses not a path at all.

Another reading of Olafur's work is possible. In the forest we are forced, again and again, to answer a particular question: this path or that? The forest becomes, in this way, a kind of physical representation of life, or at least those crises in life when a choice has to be made. Except that on these paths you can always turn back, and the exhibition you subsequently see is not made entirely different by your choice. The title of the installation seems to contain an allusion to 'The Garden of Forking Paths', a story by Argentinian writer Jorge Luis Borges, in which, somewhere in a crazy nest of narratives, the 'learned Sinologist' Stephen Albert explains to the spy Dr Yu Tsun the true meaning of a novel of that name written by the latter's ancestor: 'In all fictional works, each time a man is confronted with several alternatives, he chooses one and eliminates the others; in [this work] he chooses— simultaneously—all of them. *He creates*, in this way, diverse futures, diverse times which also proliferate and fork.' The author, Albert tells Tsun, has created 'an incomplete, but not false, image of the universe'. The universe of any single human life, though, is not infinite, any more than an art gallery is. Whichever path you choose you eventually turn up in the same place. A dead end.

Attempting art criticism of a piece I'd never experienced, I wrote to Eliasson's studio in Berlin and asked about *The Forked Forest Path*. By all accounts the studio is a magical brain factory packed with architects, particle physicists, conceptualists and light mavens. They certainly reply to emails quickly, even semi-coherent ones. Someone called Geoff sent me the original work description for *The Forked Forest Path*, which explains how boughs and branches from woodland near to where the piece is installed are, in effect, wedged into place between the floor and ceiling. He also tried to answer my questions, such as they were. Two things he wrote stood out. The first was that 'the work relates...to Olafur's beliefs that nature and culture are not separate, and that reality is something we are constantly negotiating and forming'. The second was that 'Olafur is interested in how spaces are created through movement and use (as opposed to thinking of spaces as containers)'. Or to put it another way, paths are negotiated, *formed*, within a landscape. Space is created by us rather than containing—or enclosing—us.

I followed a path into another thicket of holly. There was no way out, or so it seemed until the fallow deer hiding from me in there made a dash for it, an abrupt explosion of muscled flight ripping through the bushes in a rush of noise and movement, I myself jumping back in the other direction, my arms tensed across my chest. Having shrugged off the pull of that vegetation—a rocket moving clear of a planet's gravity—the deer sprang off and away in a silent, zigzagged celebration of space, smaller and smaller, less and less easy to follow, as it placed tree after tree between it and me, its tail the white dot bouncing along the words of its personal song.

Or mine.

CHAPTER 9

Hollow Ponds | Highams Park

To the north of the Green Man Interchange sits a boating lake next to a couple of twenty-four-hour tea huts on a stretch of land known in its entirety as Hollow Ponds. Originally a series of abandoned gravel pits, the local unemployed were paid to dig them out and join them together early in the twentieth century. This created the shallow stretch of water, nuggeted with islands, which people go rowing on to this day, dressed in everything from hijab to the cakebox *shtreimel* hats of the Hasidim, but mainly sportswear and stonewashed denim. There used to be a lido nearby too, until the 1980s, apparently quite grand, with a diving pool and art deco flourishes, all long since destroyed and buried beneath the earth. It's one of the busiest parts of the forest, always crawling with dog walkers, runners, kids trying out new bikes and young professionals taking their parents out to prove that Walthamstow and Leytonstone are not the inner city. Damon Albarn, who grew up nearby, has even serenaded it in a rather sappy song about sailing model boats, an exercise in paper-thin nostalgia.

The land is very flat and clear of trees south of the pond so that you can see right across to Snaresbrook Courts in the east and grey sky presses down on the gorse bushes below. To the north the terrain is undulant and confused, crawling with perfect oaks, their branches horizontal like the arms of Egyptian dancers, the paths running beneath the roofs they make so that when you jog through early in the morning with mist hanging over the water and magpies squawking their annoyance at you, they seem to offer a calm, quiet benediction. The bushes around here and the car park further east are renowned locations for gay cruising and dogging and if a man strolls along behind you looking as if he's forgotten something while staring at his iPhone, he's probably checking for your profile on Grindr.

To the west of the pond is a road and across that road is Whipps Cross Hospital, so that ambulances motor up and down at high speed, sirens going, day and night. In summer the road silts up as cars queue to get in and out of the car parks or bump their wheels up onto the verge. An ice cream van takes up permanent residence next to the hut where they rent the rowing boats, and often the queues for these two attractions become coterminous. A path runs along between the road and the lake, a thin, winding track just shielded from the traffic by the undergrowth, except when minor felons on community service sentences are tasked with cutting it back. For months from the late summer of 2015 onward, a cage built from four crowd-control barriers could be seen about halfway along this path, positioned down in the mud and sapling clumps at the edge of the water.

The cage marked the spot where the body of Hidir Aksakal, known as Boxer Çetin, was found that August—dismembered and placed in a bright blue Ikea bag. I say *found*, when it might be more accurate to say *bothered with*. The bag sat there for between a fortnight and a month before a dog walker, noticing the smell and the excitement of the local rats, had a look inside and rang the police. Aksakal, who at the time of his death lived in Margate, had links to Turkish crime gangs from the Green Lanes area of north London and was reputed to have been seen back in that neighbourhood in mid-August. Some locals say the bag had been there since July, a few metres south of where it was eventually *bothered with*. Aksakal had been shot and then chopped up into pieces, and whoever was tasked with getting rid of him had perhaps not devoted their full intellect to the hiding of his deconstructed corpse. By the middle of September, someone had been charged with his murder.

The truth is, there *were* Blacks in Essex, and they were there before the events in Hampshire and Berkshire which led to the introduction of the Black Act. Macaulay's *History of England* contains a reference to problems with banditry on the main forest road back in 1698:

> There indeed robbery was organised on a scale unparalleled in the kingdom since the days of Robin Hood and Little John. A fraternity of plunderers, thirty in number according to the lowest estimate, squatted, near Waltham Cross, under the shades of Epping Forest, and built themselves huts, from which they sallied forth with sword and pistol to bid passengers stand ... A warrant of the Lord Chief Justice broke up the Maroon village for a short time, but the dispersed thieves soon mustered

again, and had the impudence to bid defiance to the government in a cartel signed, it was said, with their real names. The civil power was unable to deal with this frightful evil. It was necessary that, during some time, cavalry should patrol every evening on the roads near the boundary between Middlesex and Essex.

The reference to a Maroon village (maroons being runaway slaves) is instructive, but William Addison, in his book on Epping Forest—which is rather more interested in the romance of the place than in telling you where the information it contains comes from—makes the connection even more strongly. He ties Macaulay's account to a story from 1729 in which a traveller through the forest ends up staying at a pub where the Blacks meet. Their leader calls himself 'King Orronoko, the King of the Blacks'. This is clearly a reference to *Oroonoko or The Royal Slave*, the novel by Aphra Behn, published to little attention in 1688 but turned into a much more successful play by Thomas Southerne in 1695 (incidentally the same year that John Locke was being told about robberies in the forest while passing through The Green Man inn at Leytonstone). We encountered Behn earlier as the possible—though highly improbable—offspring of Mary Wroth's offspring, neatly placing her in a parallel position to George I, the grandson of the Winter King. Her character Oroonoko is an African prince, captured and sold into slavery, who leads a revolt against his masters to free his lover, only to end up killing her (Southerne's main change to the original story was to make the lover white, a miscegenous masterstroke).

Here, then, is the clearest of equivalences drawn between the struggles of the bandits of England's forests and rebel slaves in the Americas. Though, equally, it might only be a reference to wearing blackface, much like the actor representing Oroonoko on the London stage, and indeed Mary Wroth herself in Jonson's *Masque of Blackness* one hundred years earlier. There's an instructive ambivalence at play here, a two-faced aspect which makes the issues raised hard to grasp. To blacken yourself, to present that 'very loathsome sight'—whether to hide in the dark, to embrace the saturnine or play at being African—was on some level a subversive act, though one which, today, carries rather too much baggage for an audience to look on kindly.

After the Waltham Blacks collapsed, late in 1723, the surviving members scattered, many of them choosing to lose themselves—and shake off government agents—in London. As Henry Fielding pointed out, if the alleys and lanes of the city had been built for the express purpose of concealment they couldn't have been any better: 'Upon such a view, the whole appears as a vast wood or forest, in which a thief may harbour with as great security, as wild beasts do in the deserts of Africa or Arabia.' It is then no surprise that Joseph Rose, a blacksmith and former Waltham Black, should have, at some point in the 1720s, fallen into company with a band of east London and Essex wide boys known, imaginatively, as the Essex Gang, who included among their number one Richard Turpin.

When I seriously began to think of writing this book (as opposed to seriously pretending that I might), there were two well-trodden paths I hoped to avoid. One led to William Morris, whose museum in Walthamstow was until recently a dark, dingy place full of bits of old cloth, the educational stodge palace of my nightmares. The other led to Dick Turpin, a pistol-bearing cliché, all flouncy shirts and sexy boots, a cross between Russell Brand and Mr Darcy. As you may have already realised, neither of these aversions was entirely rational. If Morris seemed too worthy, then Turpin seemed not worthy enough. Without realising it, I had been searching for a kind of median worthiness which would allow me to be flippant or overly sincere as I saw fit. Or so I thought. It turned out instead that what made me happiest was the crime of being worthy about that which I considered unworthy of my attention (and the rest of the time being unworthy about that which was worthy of my attention). My pretext came from the historian Maurice Keen, who highlighted yet another enantiodromic feature of the confusing terrain I was in. Keen points to the fact that, for the knights of Arthurian romance, the forest was somewhere unknown that they went into and got *lost*, 'the boundary of an unknown world where the laws did not run and where wicked men and strange spirits found a refuge'. For an outlaw like Robin Hood, though, it was somewhere to *hide*: 'an asylum from the tyranny of evil lords and a corrupt law'. The two worlds rarely, if ever, met, separated as they were by the gulf of class. The story of Epping Forest and its appeal to outsiders and eccentrics (taken literally to mean 'out of centre') is as much a narrative of crime and lawlessness as it is of philosophy and art. Furthermore, on a personal level, if I was using this strip of land to play out my own relationship with fear, then fear of the law, fear of *authority*, was a key part—perhaps *the* key part—of what I was examining. And crooks and thugs and bandits almost certainly had something to tell me about overcoming that. Or so I told myself anyway, succumbing to the glamour of the outlaw all over again.

•

Richard 'Dick' Turpin was born and grew up in Essex before being apprenticed in his teens to a butcher in Whitechapel, although that didn't last long. The American historian Peter Linebaugh has detailed how the collapse of guild monopolies and the inability of the City to control deals done out in the suburbs caused a massive deregulation of the meat markets in the late seventeenth and early eighteenth centuries, impoverishing many small traders and destroying the system of apprenticeship. Having learnt the basics of the craft, Turpin chucked it in, perhaps as it became clear that it was no longer a great way, or even a reliable way, to make any sort of a living. Instead, he embarked upon a life which seems to have taken in most of the iconic crimes of the eighteenth century. Indeed, this feel for the iconic—and the way in which the basics of his story attracted further romanticisation like flies to a cowpat—is probably why we remember his name now.

Turpin started off stealing cows from Plaistow marshes (the south-eastern corner of old Essex, near where the river Lea meets the Thames), then selling the meat in the back alleys and pubs of the eastern edges of the City and flogging the hides at Waltham Abbey market. Traditional accounts suggest he then joined a smuggling gang, collecting goods from boats down in the Essex marshes and transporting them to London hidden in carts. How his connection with the Essex Gang began is unclear, but with his contacts in the meat trade it's likely that he was already familiar with poachers, either butchering stolen venison or selling it on their behalf. In addition, as I've already mentioned, this band of deer thieves included a former Waltham Black, Joseph Rose, who may have had smuggling connections of his own. In the early 1730s, these and other gangs were causing the authorities considerable trouble, with the Verderers of Epping Forest writing to the government in 1731 to complain that 'diverse disorderly and idle persons have of late … killed, wounded and carried off several of His Majesty's red and fallow deer within said forest of Waltham'. In response, a reward of ten pounds was announced for information leading to the men's capture.

After the Essex Gang carried out a particularly brutal raid on Enfield Chase in 1733, the reward was upped to fifty pounds. The gang decided it was time to give deer-stealing a miss, turning their attention instead to robbing people in their houses. Twice in October 1734 and twice again in December, they barged their way into shops and homes in Woodford, Chingford and Barking and forced the residents, with threats and violence, to give up everything of value. This spree culminated with an attack on Hainault Lodge, the house of William Mason, a Keeper of Epping Forest, in revenge for his investigations of deer-thieving. Somewhere between six and fifteen men turned up, all with faces blacked, beat up Mason and fired shots at one of his servants as he tried to escape to call for help.

January saw the gang working further afield, but in February they were back in Loughton, where they threatened to place an elderly widow on her own fire if she didn't tell them where her money was: 'God damn your Blood, you old Bitch, if you won't tell us I'll set your Bare Arse on the Grate.' Three days later, they attacked the farm of Joseph Lawrence in Edgware. They smashed up the house, Turpin beat Lawrence across the bare buttocks until they were black and blue and a fellow gang member, Sam Gregory, raped a maid at gunpoint. You can see why E. P. Thompson didn't want to draw a connection between these actions and those of his *social bandits* who only ten years before had, on the other side of London, carried out their actions 'to see that the rich did not insult or oppress the poor'.

Over the next two months, all of the crew were caught except for Turpin and one other gang member, Thomas Rowden. Lacking the manpower for house raids, the pair switched their attention to highway robbery—first at Mile End and Epping Forest, later over towards Wandsworth and Barnes. This crime wave—basically a collection of muggings glorified after the fact—continued until the autumn, when the pair separated, with Rowden heading to Gloucester and Turpin apparently leaving the country to hide out in Holland for a while. Rowden was later caught and transported for forging money, but Turpin was now set on his path. By 1737 he was back in England and once again robbing out on the roads.

At some point he met another highwayman, Tom King. It has been claimed that Turpin tried to hold up King, a kind of meta-robbery which caused the pair much crooked amusement, though this is almost certainly an invention. Whatever the case, the two of them began to work together, and perhaps King added some strategic smarts to the partnership. As part of their planning, they decided to make a hideout in the forest, large enough to accommodate their horses, from where they could see two key roads and attack anyone who passed. From this base the duo raided traffic up and down thoroughfares as far away as Suffolk. Turpin's Cave became a key part of the highwayman's legend, a secret forest hideaway to which he could return, a kid's den made real: 'In this Cave they liv'd, eat, drank and lay; Turpin's Wife supplied them with Victuals, and frequently stay'd there all Night.'

On 29 April, fleeing from an ambush in Hertfordshire, Turpin stole a racehorse near to The Green Man pub in Leytonstone, these days situated next to the roundabout we visited earlier but back then deep in the forest. Although the landlord of this pub, Richard Bayes, would write an account of Turpin's life and execution a few years later, at the time he was widely believed to be in league with crooks of various sorts, the highwaymen among them. It's what landlords have always done, and was one of the key reasons for attempts to shut down hostelries in the forest. This complicity is borne out by Bayes's account of what followed. On being told of the theft, he says, he immediately knew that Turpin was responsible and also where to locate him in London—the Red Lion in Whitechapel. Going there, Bayes found the stolen racehorse in the stables and was examining it when Tom King appeared. King, he claims, drew a pistol, placed it against Bayes's chest and fired, only for it to *flash in the pan*. The two men were struggling when Turpin appeared. King told Turpin to shoot Bayes. Turpin accidentally shot King, twice, killing him, after which he fled. There's really no such thing as a secret hideaway, and within a few days of the incident, Turpin was seen and tracked back to his *cave* by Thomas Morris, a servant of one of the Forest Keepers. He tried to apprehend the thief and, for his troubles, was also shot and killed by the highwayman.

I decided on another expedition in futility. There's something very calming about looking for something you know you won't find. I'm not sure if there's a word for that. Desearching, perhaps? I decided to go and desearch for Turpin's cave. It was a pointless, hopeless task. For one thing, the land around this area of the forest—west of Loughton, up on the humped

spine where the trees are thickest and most numerous—
is a Lower Bagshot Bed, a mixture of sand, loam, pipeclay and
flints that originally formed under shallow water sometime
back in the Upper Eocene era, over thirty-five million years
ago. While it's salutary to think of this whole spit of land as a
massive sandbank in some warm prehistoric estuary, it's not
the kind of soil that supports a walk-in cave. My best guess is
that Turpin and King's hideout was more like a deep pit dug
back into a bank with some sort of rudimentary roof added,
the whole disguised with branches and bracken and holly
and so on. What I was looking for now was a steep slope in
which could be seen an overgrown hollow that may never
have existed.

The earliest accounts of its location placed the *cave* near
to Loughton Camp, the southernmost of the two Iron Age
hill forts to be found in the forest. The camp lies more or less
directly west of Baldwins Pond—the small lake near to where
both the Epsteins and Ken Campbell lived—but the usual
route there is to follow the twists of Loughton Brook up from
Staples Pond, which, in turn, is more or less straight up the
hill from Loughton's underground station. Where the brook
crosses one of the major paths—Green Ride—you cross too,
following the whisper of a path through grass and cutting
up the hillside through the trees. It's testament either to the
uniquely disorienting qualities of the forest or my own lack of
route-finding nous that even from this point I have managed
to blunder past its banked-earth perimeter, enclosing an area
the size of four football pitches.

In summer, when the tall beeches are fully foliate, the light
here is green and suitably submarine, and flocks of small birds
move through the branches of the younger trees and bushes

223

like tropical fish dashing between coral outcrops. In January things are much more stripped back, what light there is bouncing from the browns of mud and dead leaves. I worked my way round the fort's south-western flank, following the stream. On this perambulation I found the triangular box for a pre-packaged sandwich. A butterfly, its wings eyed on top and black beneath, made a ragged flight away from me. An old plank with rusty nails had been leant up against a tree. A buzzard took off above me. There was nothing even like a cave, and I wasn't really looking. It was hard to tell if I was dowsing for ghost traces or just bunking off. Actually, it wasn't. When I reached the very top of the stream, where it disappeared into a narrow tunnel beneath the road, I sat on a fallen hornbeam to check the map and eat chocolate. I plucked at something down near the exposed roots. When I pulled it away it almost crumbled in my hands—an iron chain so decayed it looked more like bark or twigs grown into interlocking loops. Even the man-made is organic. It depends how close you stand.

I worked my way back round, higher up the slope, strolled up to the embankment, wandered about kicking Bagshot Bed soil from the roots of other toppled trees. A different account placed Turpin's hideout to the north of the camp, so I tried there. It was mainly flat with a lunar landscape of abandoned gravel pits, an adventure playground for the mountain bikers who are meant to keep off the hill forts but clearly don't. Beyond that, the holly got so thick I had to skirt round it, and if the cave was secreted somewhere behind those bushes it would've been a soggy, sorry sort of place. I came out on to a wide, marshy plain, my boots sinking into water, a man in old sportswear having a piss right out in the middle, his muzzled husky devoting too much attention to me.

Before I went to the next possible cave location I thought I'd call in on the Skull Tree, as it sits quite close to Loughton Camp. I stumbled north-north-east and saw a pair of what were either woodpeckers or extremely antisocial thrushes. I saw a blue pen lid, a food container, an empty Coke bottle nesting in a tree, a discarded tin of coconut water. You might have thought I'd collect the litter, but it seems natural here, as much a part of the place as moss or leaves. I had decided a few months back that the Skull Tree's skull in fact represented Ken Campbell with a plunger stuck to his head—one of his favourite party pieces and a key motif in *Jamais Vu*—and had rushed back to check the initials carved next to it, disappointed to find out they were an L and an E. The beech which had fallen nearby was all dead now, giving the aspect a slightly melancholy edge. If the skull represented anything of Ken's legacy, it was in its own enantiodromic way. Today it seemed to have grown a beard, so that it looked like Isambard Kingdom Brunel in a novelty stovepipe hat, or a character from Dickens. I couldn't think how this related to any search, re- or de-, but I was happy sitting on my perch eating my sandwiches. Now the leaves had gone I could see the cars buzzing past on the road less than a hundred metres away.

From there, I headed across to the Kings Oak pub at High Beach, following the path I'd originally used to relocate the Skull Tree, crossing the main road and then cutting round the back of the Epping Forest Visitor Centre, which—like an old widower—seems to keep irregular hours these days. The trees break at the Kings Oak so that you can see right down the western side of the hill and out over Waltham Abbey and the Lea Valley, everything smudged and elusive. The pub itself is a cavernous hotel fallen on hard touristic times, the management of which is trying to compensate by offering some kind of Ibiza pool-party experience at the unheated mini-lido it has out back, one of those glaring reality chasms opening up between intent and actuality. Having stopped for tea and excellent bread pudding at the counter next door, I continued down Wellington Hill, looking for my next destination. Up until the 1960s or 1970s there was a pub here, on the western flank of High Beach, called Turpin's Cave, sitting right next to some sort of giant hole: 'This excavation, or what is claimed to be the one, is still pointed out,' Addison tells us. 'The cave itself is twelve feet long, seven feet high. In Turpin's day it was probably larger. Inside the inn are an old cutlass, with rusty iron scabbard, the head of an old pickaxe, old horse-shoes and horse-bits, two large padlocks, a pair of rusty handcuffs, a flintlock pistol, and similar objects, which the visitor is assured were discovered in the cave.' It took me a while to find the spot, the pub having been replaced by a large residential property and the pit presumably filled. It was in the wrong place, anyway, more of a failed marketing ploy than a site of historical interest.

After the murder of Thomas Morris, Dick Turpin headed north. He no longer had a hideout, the reward for information leading to his capture had been widely advertised at two hundred pounds and too many people in London and Essex knew who he was. He travelled first to Lincolnshire and then on to Brough in Yorkshire, arriving with a new name and persona: John Palmer, gentleman. Perhaps he never said he was a gentleman. Perhaps he didn't need to. The fact that he didn't work and spent his time going hunting with the other gentlemen of the neighbourhood was clue enough. Every few months he would vanish off to visit friends in Lincolnshire.

When he came back he brought horses with him to be sold. No one seems to have questioned this, and why would they? His actions were guarded by his presumed class.

You have to ask, though, whether people really believed Turpin was a gentleman, or if they simply didn't care so long as he wasn't robbing them. It's hard to disentangle this part of the story from the myth of the highwayman-as-gentleman, removing his glove to show that his hand is pale and clean and unsullied by manual labour, a tale which crops up in different forms in relation to Robin Hood, to the Waltham Blacks and to Claude Duval dancing the courante with a victim's wife during a hold-up. It's worth noting the way that stories of the highwayman—and indeed highway robbery itself—grew at the time that Enclosure was accelerating the development of capitalism and, with it, a new set of social relations. It was the first time in hundreds of years, perhaps the first time ever, that you couldn't be quite sure who was and was not a gentleman. The first time outside of the forest, anyway.

The other possibility is that everyone was scared of John Palmer. He stayed in Yorkshire for the best part of two years before he made a mistake. Returning from a disappointing day's hunting, he saw a cock belonging to his landlord and shot it, presumably more to entertain his gentleman companions than for the sport. A neighbour of his, servant to the landlord, saw him do this and came out to remonstrate. Palmer told the man, Hall, that 'If he would only stay whilst he charged his Piece, he would shoot him too'. Hall scuttled off to tell the landlord. That was how Palmer was arrested.

At this point Turpin made another error. Bound over to keep the peace, he refused to pay the bond which would've secured his release and so was moved to the House of

Correction at Beverley and held there. Once he was safely behind bars, the locals woke from their trance and raised concerns about where the horses they'd been buying and swapping with Palmer were coming from. Word was sent to Long Sutton, the town he claimed to have lived in previously in Lincolnshire. A letter came back saying that he was wanted for sheep-stealing there and that there were accusations of horse theft too. Palmer, now a more notable prisoner than anyone had allowed for, was taken to York Castle.

He might still have got away with transportation at that point, but he was betrayed by his literacy, a salutary lesson to all writers. He sent a letter to his brother-in-law back in the village of Hempstead in Essex, where he grew up. The brother-in-law wasn't expecting it and so refused to pay the postage. Another man, a Mr Smith, who had taught the young Turpin to write, saw the letter and, it's claimed, recognised the hand-writing. With his own hands shaking, presumably dreaming of the two hundred pounds that might be coming his way, Smith paid for the delivery of this missive and took it to a magistrate, who opened it. In it, John Palmer told his 'brother' of his imprisonment and appealed for him to get character witnesses together. Smith, sure now of the writing, or just itchy for the money, resolved to go to York and, on being shown Palmer, confirmed that he was in fact the noted highwayman Richard Turpin. And Turpin must have felt a tightness round his throat, his legs swinging lightly in the breeze.

The contemporary view of Turpin, drawn in large part from James Sharpe's excellent book *Dick Turpin: The Myth of the English Highwayman*, is that he was a violent hood—not even a particularly effective one at that—only saved from obscurity by William Harrison Ainsworth, when he made him a character in his 1834 potboiler, *Rookwood*. The novel represented a last throw of the dice for Ainsworth himself, who was struggling with debt and a collapsing marriage, and was also a lawyer and so desperate to do something—*anything*—else. *Rookwood* is a gigantic mess of a book, a Gothic romance set in Yorkshire packed with murder ballads and songs so awful that even Thomas Pynchon would cut them, about as overwrought and silly as a 1980s Venezuelan soap opera but with none of the tanning or special effects. It has forests and poachers and Gypsies, plus family curses, incest, gunfights, skeleton hands, nice-but-dim leads and sulky villains. It sounds tremendous put like that, but it's not. It's awful. It contains lines as shockingly bad as, 'Nor had he to wait long before its invigorating effects were instantaneous'. It has no characterisation as such, but no real plot either. Or rather, it has a plot but it's convoluted and ridiculous. The more I tell you how bad this book is the more I worry you might think I'm recommending it. I'm really not. It's not even clear what Dick Turpin is doing in there, except that the model for the book was the kind of gothic romance Ann Radcliffe had popularised and, in relocating the style from Italy to England, Ainsworth needed a replacement for the brigands. An alliance between Turpin and the local Gypsies dealt with that.

Ainsworth takes the story told by pub-landlord-turned-true-crime-author Richard Bayes almost a hundred years earlier and ropes it into his narrative, changing what needs changing, enhancing what's too boring, monkeying around with it, compressing events and, in the process, reinvigorating or inventing every highwayman trope going. But it's the story he added to Bayes's account which resonated most with the public, often being published on its own as a pamphlet, freed from any need to fit, and this was Turpin's overnight flit from London to York on his trusty steed, Black Bess. Based on an old tale about another highwayman, the idea is of a villain completing a seemingly impossible journey in order to generate a sturdy alibi.

In *Rookwood*, though, Turpin is chased all the way and does it for the sport and immortal glory rather than anything as utilitarian as escaping the gallows. He's pretty fatalistic about all the death stuff, seeing it—much like a young rock star contemplating his imminent overdose—as very much part of the job. And in the process he works poor Bess to death. It's a florid, fervid, sexualised episode. His partner in crime, gentleman-gone-bad Tom King, has just been betrayed by his mixed-race beloved, but the pure-bred Bess—'Is she not beautiful? Behold her paces! How gracefully she moves! She is off!'—keeps going and going till she can give no more: 'There was a dreadful gasp—a parting moan—a snort; her eye gazed for an instant, upon her master…a shiver ran through her frame. Her heart had burst.'

Ainsworth made much, too, of the writing of the episode, 'a feat' in which he claims to have bashed out one hundred pages in one sitting lasting a day and a night:

Well do I remember the fever into which I was thrown during the time of composition. My pen literally scoured over the pages. So thoroughly did I identify myself with the flying highwayman, that, once started, I found it impossible to halt. Animated by kindred enthusiasm, I cleared every obstacle in my path…With him I shouted, sang, laughed, exulted, wept. Nor did I retire to rest till, in my imagination, I heard the bell of York Minster toll forth the knell of poor Black Bess.

My own writerly emotions are held much more tightly in check, though I do emit the occasional strangled whine due to bad posture or too much coffee.

It's been argued, and persuasively, that it is *speed* which caught the imagination of the reading public. The journey— some one hundred and fifty miles—was clearly impossible to accomplish on horseback in a night, but this was just as steam trains were being introduced and so what the story in fact plays on is a new fascination with previously unattainable velocities. There's also a link to the myth fostered by Jack Kerouac who, asked how long it had taken him to write *On the Road*, replied, 'Three weeks.' Both books display the same interest in low lifes and outlaws, Gypsies and tramps, and the spontaneity of the writing is meant to reflect the spontaneity of the participants while at the same time contrasting it with the steady dullness of bourgeois *straight* society.

Indeed, after writing a more conventional historical romance, *Crichton*, Ainsworth returned to outlawry in 1839 with *Jack Sheppard*, about the thief and famous prison-breaker of the same name. Taken aback by how quickly these books were transferred to the stage and hence consumed by the illiterate working classes, the literary world began a moral crusade against what were termed *Newgate novels*, after London's most famous prison and, from 1783, execution site (and the scene of the real Jack Sheppard's two most famous escapes). 'All the Chartists in the world,' wrote Mary Russell Mitford, 'are less dangerous than this nightmare of a book.' The situation was to become still more inflamed. On 6 May

1840, Lord William Russell was murdered in his bed by his French valet, François Benjamin Courvoisier, who was tried, sentenced and hanged for the crime that July. The papers went into overdrive, claiming that the valet had read *Jack Sheppard* in the days leading up to the murder and that this was what had provoked his actions: 'If ever there was a publication that deserved to be burned by the hands of the common hangman it is *Jack Sheppard*.'

While it's always easy to dismiss a moral panic—and we have newspapers devoted to creating them to this very day—it seems there was some basis to these fears. In *The Condition of the Working Class in England* (1845), Friedrich Engels quotes a certain R. H. Horne, at the time working for a Royal Commission into the employment of children in factories and mines. 'Some of the children have never heard the name of Her Majesty,' Horne had written in his report on the children of Wolverhampton, 'nor such names as Wellington, Nelson, Buonaparte. But…among all those who had never heard such names as St Paul, Moses, Solomon etc, there was a general knowledge of the character and course of life of Dick Turpin, the highwayman, and more particularly, Jack Sheppard.'

How do you write about crime without glorifying or romanticising that crime, turning every murderous thug into a dashing rogue? There's no need to get bogged down in essentialist nonsense, but it's true to say that one of the things that distinguishes humans from all other life we have so far encountered is not our ability to communicate—all kinds of animals do that—but our ability to tell stories. And further, it's possible to argue that when we tell stories we glamorise, because we give shape to the shapelessness and senselessness of life and, in doing so, make it somehow more attractive, more compelling, than the shapelessness and senselessness of our own existence (which is why we are constantly telling stories about ourselves, both to other people and as an internal accompaniment to our actions). *Glamour* shares its roots with *grammar*, and both imply a kind of magic or occult learning. Stories, though, are enantiodromic. They can be shaped to be cautionary or liberatory, subversive or conservative, and it's not always clear which is which, let alone which is preferable.

B ack down in the forest, two and a half centuries later, we find Turpin's shadow. Harry Roberts is no relation of Mick Roberts, though both of them were involved on the fringes of London's gang culture in the 1960s, and the former provides a twisted counter-narrative to that of Old Mick the eco-warrior. Towards the end of 2014 Harry Roberts was released from prison at the age of seventy-eight. He had served forty-eight years behind bars. On the night his release was announced I heard an anonymous policeman on the radio—not the best and brightest member of the force, judging by his verbal skills—state that not only should Roberts not have been given parole, but that he should be executed for his crime. That crime was the shooting of three policemen in 1966, when two of them made the mistake of approaching the vehicle he was sitting in with Jack Witney and John Duddy, near to the walls of Wormwood Scrubs Prison. The officers were only checking up on an expired tax disc, but Roberts pulled out a Luger and shot them before Duddy killed the third policeman as he tried to escape.

Roberts's accomplices were caught very quickly, but he himself proved to be more elusive. Despite a massive man-hunt, he stayed at liberty for ninety-six days, combining the army training he'd received killing communists in Kenya and Malaya with a thorough knowledge of the forest running up from his original stomping ground of Wanstead to Bishop's Stortford, where his family had gone on camping holidays when he was a kid. Finally captured hiding under straw in a barn near the latter town, Roberts was brought to justice, and he and his associates were tried and given life sentences with thirty-year minimum terms. John Duddy would die in Parkhurst Prison in 1981, while Witney was released ten years later, only to be beaten to death by his flatmate before the turn of the millennium. Roberts, however, stayed inside, both a bogeyman and, increasingly, a source of concern to liberals and prison reformers. Stories circulated of him making pies decorated with crusted images of cop-killing, a feat which should surely have won him a berth on *Celebrity Bake Off* (or a spin-off involving famous crooks baking cakes containing files). On the other hand, he was way past his minimum term and only a succession of Home Secretaries was blocking his release, ostensibly to keep the Police Federation happy.

This confusion reached its peak in the early years of this century, when the Home Secretary, David Blunkett, refused to grant Roberts parole, but said that the information his refusal was based upon must be kept secret not just from the general public, but even from Roberts and his legal team, a decision which caused a certain amount of consternation. Former hostage Terry Waite was among those who spoke up, pointing out that 'the principles of fairness and justice should be applied equally in a democratic society, however heinous the crime or the criminal'.

But before the decade was out, the *Mail on Sunday* claimed that Roberts had been refused parole again because of a campaign of harassment against Joan Cartwright, the owner of an animal sanctuary he had been working at when on day release from an open prison in Derbyshire. On being turned down for parole in 2001—the case Blunkett had ruled on— Roberts decided, claimed the newspaper, that it must be because of evidence given by Cartwright about his ongoing criminal activities. In the article she stated that for the next four years he called her almost daily to threaten her, saying that anyone who testified against him would be 'torn limb from limb'. In addition, Cartwright believed that associates of Roberts were responsible for regularly killing and maiming animals at the sanctuary, including strangling a peacock and electrocuting her cat.

If the story is true, though, how is it possible that only five years later Roberts was released? It doesn't add up, even without the cartoon animal executions, as if the need for an evil villain is its only imperative. And here's the thing, here's where Roberts really links back to Turpin, beyond the excitement of that long hunt and his purportedly impressive evasions— in the way he fills a need, a particular kind of hunger for a particular kind of archetype. Because while the mainstream narrative is of a brutal, psychotic killer who should never be released, football supporters still, to this day, take up the occasional chant of *Harry Roberts is our friend / Is our friend / Is our friend / Harry Roberts is our friend / He kills coppers.*

When I told people I know locally that I was working on a book about the forest, murder was often the first thing they thought of, telling me about the time a severed head was found out on Wanstead Flats behind their house, or asking if I'd be covering gangsters and dead drug mules and so on. Murder—usually vicious, pointless, brutal murder—is part of the DNA of these woods, the evidence collected at the scene, so to speak, and we seem to take a strange kind of pride in that fact, as if we are made more real, braver, grittier, by our proximity to it. Rumours of the dumped corpses of failed back-room abortions. Boxer Çetin in bits in a bag. A man with his hand hacked off at the wrist outside a club in Loughton. Babies chopped up and chewed in satanic rituals at the Gypsy Stone. In telling its story, we invest murder with meaning. But is to give it meaning, even the most basic of meaning, already to burnish it?

Poverty breeds desperation; it breeds crime and mental illness. Penny Rimbaud suggested to me that the poorest neighbourhoods in Western European cities tend to be on their eastern side, because that's where prevailing westerly winds push the pollution. A wasteland lying on the eastern fringes of London is, then, likely to act as a spillover or reservoir for the crime that city generates. If Fielding saw the forest in the city, then doesn't it follow that the city is in the forest? And this *explanation* is almost as trite as the acts themselves, so perhaps it's better to offer no explanation at all. In which case there's nothing to learn, except that a hiding place is morally neutral and, if it can be used to shelter artists and nonconformists, it can also be used to hide murderers and corpses. I took my son out and showed him the Skull Tree one Sunday. He didn't find it amusing or diverting. He found it creepy. He scanned the trees around us as if we were being watched. He saw the skull less as an interesting puzzle and more as a warning, like a Keep Out sign or the symbol on a bottle of poison. Murder is a crisis resolved badly—resolved calamitously—from which can flow only more crisis, an intensification rather than a longed-for release. Murder is banal and nasty and any description of it partakes of this banality and nastiness, pretends to nothing more than pornography. Murder is a meaningless list soaked in sadness and trauma.

The thirty-one-year-old woman, sent to buy twelve hundred pounds worth of drugs and returning with a bag of parsley, was stripped and beaten to death with a pool cue by the men she was running the errand for. The young girl and boy, eleven and twelve respectively, were raped and then strangled in a den the killer had built or commandeered near Lippits Hill. After the man was murdered and dismembered, his body parts were spread out over Leicestershire and Hertfordshire—except for his hands, which, according to the killer's drunken girlfriend, were buried somewhere near to Sewardstone. The body of an unnamed woman was found close to Bury Road in Chingford, but the police weren't treating her death as suspicious. The forty-year-old builder was lured to the forest, stabbed to death and then burnt by his lover, his lover's brother and other men who were never caught. The thin, decomposing corpse of a thirty- or forty-year-old man was uncovered, his money belt containing British and Spanish currency. The bloodstained clothes of the four-year-old were found stuffed into a hollow of a tree six months after she vanished. The police knew the deceased had been married because of the wedding ring, but the body was so decayed they couldn't ascertain whether he was someone's husband or she was someone's wife. The couple were shot while sitting in his car, leading to speculation that they were the victims of a contract killing. The female provider of 'discreet sexual services', also in a car, had three crossbow bolts buried in her head. On leave from the RAF just after the war, he drove his girlfriend up to the woods one night, but a man with a gun opened the car door, dragged him out and shot him three times. The young Pole, twenty-five, was found hanging from a tree. The body of the former Mayor of Hackney was at first mistaken for a pile of rubbish at the roadside, after he was kidnapped and had a heart attack during a botched robbery. An associate of gangsters and headcases, the man was stabbed during a brawl at an infamous nightclub near Chigwell and died in the back of a cab on the way to Whipps Cross Hospital. The drug smuggler was beaten, scalded with boiling water and then suffocated before his body was dumped near to Rangers Road by an associate who believed he'd been tricked out of his share. Fourteen years old and living homeless in London, the boy was gagged and repeatedly raped by a group of men until he died of internal injuries and asphyxiation, his naked body left near Ongar wrapped in a thin blanket. The man went to a remote part of Highams Park to have sex with a younger woman, but her boyfriend was waiting and after he had beaten him to death with a spanner, they took his money and watch and used the funds to get married. Fifteen, walking home from a party, she was strangled and left in the brambles less than a hundred metres from home.

And all this is part of the forest too.

CHAPTER 10

Crown Hill | The Cloven Tree

I f you think of Epping Forest as the edge or shadow of London, then Crown Hill is the edge or shadow of Epping Forest. It's the road that leads out from the centre of the protected woodland to Upshire, where some of the first motorway protests began. The efficacy of those protests is demonstrated when you turn the corner as you reach the bottom of the hill and see the gigantic concrete stanchions of an M25 bridge lurking there, the ground beneath thin sand, as if prepared for the filming of a 1970s BBC science fiction series. I didn't get that far on this particular day, stopping instead to stare at the footpath I'd hoped to follow. The stile was broken and the path—which ran in a narrow, fenced-off strip between a house and the *storage centre*—was so overgrown it looked more like a feudal concession for nettle cultivation than a right of way.

I'd already had to slow my approach as I walked down the road while a man with joggers falling off his arse bent down and chained the gate of the *storage centre* shut, looking back over his shoulder at me with a blank mixture of contempt and suspicion, before getting into his car and motoring off. It was growing late, and what light the clouds let through was starting to follow him away. I couldn't be bothered to go back and there was no way to go round. I edged over the wobbly fence and picked my way through the vegetation. It was very quiet, far too quiet, all things considered. I waited for the promised guard dogs to bark, to come chasing towards me, hopefully to be yanked back, half strangled, by sturdy metal chains. The yard of the *storage centre* to my left was full of junk—old fire extinguishers and bits of car, skips overflowing with assorted industrial crap, Calor gas containers, broken chairs. A long warehouse with a corrugated iron roof made me think of abattoirs, or at least of amateur abattoirs, that's to say, of the kind of place where serial killers operate in crime dramas. It was the stillness, the lifeless quality, which got to me. I forced my way through even thicker nettles, climbed over another rickety stile and made my escape along the edge of a ploughed field.

I never saw what I'd hoped to glimpse there. But, as I headed south and south-west to where I could get my train home, I came across lots of other man-made hideouts, secret bases, caves, teepees and forts. The forest is full of huts, most of them temporary, unfinished or to be completed only in the imagination of their builders, a contingent architecture. Sometimes a low branch is used as the spine for them. Often it's a bough that's broken off which is then lifted and butted up against a living tree at one end. Or a youngish tree has been blown over and come to rest against one of its neighbours. It doesn't matter, as long as it serves as a roof beam. Branches are piled up on each side of this central beam to varying degrees of thickness and, in doing so, an interior is created, low and dark and damp, a space for crouching and picturing yourself a hunter or a warrior, a chief, a robber, a rogue, an outcast. Occasionally, dirt and leaves are added to make the roof more impenetrable but this is unnecessary, really. No one is trying to make a genuine shelter.

My own children used to have one. They found a hollow covered by bushes beneath the roots of a fallen tree not that far from where I was now walking. We went back there a few times so they could make minor improvements, though most of the game was in their heads—planning what they would bring back the next time they came, which they never did. I still remember their horror on arriving there one time to find that some other children had left a tin (I forget what it contained) in *their den*. Some of the continuing fascination with the story of Dick Turpin—a mugger with an excellent PR department—is the appeal of his den out in the forest, where 'they liv'd, eat, drank and lay... and frequently stay'd there all Night'. That and the number of times he seems to have jumped out of windows to evade justice: what Peter Linebaugh calls a myth of excarceration, the escape we all dream of.

But are these two impulses—to build the hut and to escape—compatible? Turpin, after all, had to kill a man when he was finally cornered at his cave, bamboozled, perhaps, by the lack of fenestration to fling himself through. Writing about Jules Verne, Roland Barthes notes the author's 'delight in the finite, which one also finds in children's passion for huts and tents: to enclose oneself and to settle'. It's an interesting idea, counterintuitive, as of course Barthes intended it to be, foregrounding the hut as a kind of enclosure, which at the same time it self-evidently is. Verne, he goes on to argue, 'in no way sought to enlarge the world by romantic ways of escape or mystical plans to reach the infinite: he constantly sought to shrink it…to reduce it to a known and enclosed space…The basic activity in Jules Verne…is unquestionably that of appropriation.'

I have a feeling, though, that Barthes didn't spend too much time in huts. It's not how Parisian intellectuals get their kicks. Leave that stuff to the Germanics, to Wittgenstein or Heidegger: 'Solitude has the peculiar and original power of not isolating us but projecting our whole existence out into the vast nearness of the presence of all things.' In my admittedly limited experience, everything Heidegger ever wrote reads a little like a pastiche of how you'd expect Heidegger to write—which is just the Beingness of the Becomingso of the vast Blah of my own Shallowness—but there is something in that formulation, something which relates to the specific type of enclosure the hut provides. It is less a barrier to everything outside than a permeable membrane.

We could turn to Thoreau, of course, the ultimate huttite, but let's stay closer to home. Roger Deakin, the writer and environmentalist, used to sleep out in a shepherd's hut in his garden as often as he could. 'Why do I sleep outdoors?' he asks. 'Because of the sound of the random dripping of rain off the maples or ash trees over the roof of the railway wagon, or the hopping of a bird on the wet felt roof, or the percussion of a twig against the steel stove-chimney. Out there, I hear the yawn of the wind in the trees…I feel in touch with the elements in a way I never do indoors.' On this reading, the hut becomes not a way to shut out the world, but a practical way to be closer to it, to let it in. Perhaps, like John Clare— repulsed by Enclosure but titillated by the opportunities for transgression it presented to him—the excitement of the hut lies exactly in the tension between these two opposing desires, the urge to shut out and the urge to *be out*. I should know. I spent a week reading about tea ceremonies, about ritual, about architecture, about hermits and monks and philosophers. I had managed to skirt into the issue of lawlessness and then skip straight back out again, without even noticing, my path taking a sudden scoot off into a different terrain. I wanted to be in it and also to keep it out, crouched in my makeshift shelter, a study in repression.

I had gone to grab a look at the *storage centre* because of exactly this obsession. I decided quite early on that I needed a base out in the forest, somewhere to come and write and drink tea and be surrounded by trees. An escape pod, a cocoon, a man-womb, a scribble-shack. Looking online, I found that someone was renting out as office space the kind of prefab huts you get on building sites and, although they were somewhat lacking in the basic aesthetic requirements you'd want of a woodland hovel, they were available. The only problem, beyond their eye-watering weekly cost, was the location. Somewhere down at that *storage centre*, backing up on to the enthusiast's abattoir, inside which men in saggy joggers took turns to work the handheld sausage machine like leftovers from Grimm. In truth they were probably just fixing cars. It didn't matter. I'd already placed the *storage centre*, decided where it belonged. It was right in the middle of my worst nightmares.

There is another hut in Epping Forest, not available for rent, which you can only go to see by special arrangement. It's hidden away at The Warren, up on the Epping Road between Loughton and Chingford. Back in the eighteenth century The Warren was a high-class forest retreat called The Reindeer

Inn, specialising in rabbit pie. It's now the forest headquarters for the City of London Corporation. If Crown Hill—in my mind at least—represents the illicit, shady side of the forest, then The Warren is its opposite, all clean lines, whitewash and bureaucracy. The villa itself is symmetrical and simple, its windows looking north because, as the landscape designer Humphry Repton pointed out when asked to make a report on the building, most of the inn's guests would have come in summer, and north-facing rooms would've been the coolest. The hut sits somewhere up at the back, not just behind the elegant mansion that accommodates the people who run Epping Forest, but behind some outbuildings and, eventually, behind a petrol pump. It's a rather lovely structure if you're into sheds, the opposite of a soulless prefab, an 'L' of two equal lengths, the roof coming down higgledy-piggledy from the back wall in runnels of mossy old clay tiles, the front wall's large windows running from around knee height up to head height, where they meet the guttering. The glass makes it very light inside, though it would've been a nightmare to heat. Originally, one side was open along its front, a cloister which was walled in at some point in the last twenty or thirty years to make a classroom for inner-city school visitors, an idea since abandoned. The place is now used to store tools and waders, old kettles and buckets and various bits of junk. It smells of shed—a mixture of damp wood and creosote and something indefinable, something romantic and nostalgic. I'm not sure that all huts smell like this, but all the great ones do.

The hut was dragged here sometime around 1930 when the land it stood on over towards Chingford was sold to the Corporation and became part of the forest. Before that it sat on Pole Hill, so named because of a marker placed there in 1824, right on the Meridian line, to help the telescope at the Royal Observatory, Greenwich, align itself with true North. It's a short walk there from Chingford station. You follow the road opposite past the Station House pub, whose small windows give it an aggressive squint, then cross the parody of suburbia which is The Drive, where the houses seem actively to *mock* Tudor. Carrying on up, the buildings shrink a little, become pebble-dashed, as if you're returning to the 1950s, then the road ends and you step into a small green tunnel through the trees to the summit. The first time I was up there a group of mountain bikers stood around, their bikes between their legs, discussing whether a certain descent was a piece of piss or not while a little girl threw an epic tantrum and told her dad—and the world, indeed the universe—that she hated him. It's only a small hill, really, but the views back over London have a certain understated majesty, somehow rose-tinted, turning the city into a view Tolkien would admire.

The hut which now rested round the back of The Warren was called Cloisters and it was built by the 'Welsh-American "metaphysician",' Vyvyan Richards. Richards was a teacher at the nearby private school, Bancroft's, and had been since he graduated from Oxford. He ran the school's scout troop, and apparently it was those boys who helped him build it. I was shown the hut by Judy Adams, chair of the Epping Forest Centenary Trust, and her husband Peter, who is a Verderer. I couldn't place Judy's accent—perhaps something antipodean in there, flattened out by years in England, and then something else which made me think of Gloucester, a place I don't think I've ever visited. Peter was donnish, with the sort of right eyebrow which lets you know when he thinks you've said something stupid. Peter told me in no uncertain terms that the building we were looking at should be known as the Richards Hut. It's not. It's known as the Lawrence Hut.

T. E. Lawrence continues to exert a strange fascination seemingly far in excess of his achievements. For some time this celebrity was driven by the movie *Lawrence of Arabia*, though this has itself sunk back into the cultural mud, remembered only for starring Peter O'Toole rather than for anything that happens therein. All the same, people know the name, and his acolytes still celebrate him. Compared to his heyday, though, Lawrence is now obscure. He was, from shortly after the First World War until the late 1930s at the very least, a star. Churchill and E. M. Forster both attended his funeral. He knew and corresponded with just about everyone who was anyone—Robert Graves, George Bernard Shaw, Noël Coward, Thomas Hardy, John Buchan, Joseph Conrad and many more. Jacob Epstein's last major stone work, *Lazarus*, from the year of his wife's death, 1947, was taken to be a memorial to Lawrence. More personally, my father was named after him when he was born in 1938. But here's the thing. A few years later my future grandparents—overnight, with no warning—began to call him by his middle name and when he went to school on the following Monday, the teacher had been told to as well, so that Lawrence the boy growing up in east London just *vanished*, a prefiguring of Lawrence the man's fate. Perhaps that's apt for a person who said of himself, 'there was a craving to be famous; and a horror of being known to like being known'. There were two very distinct sides to Lawrence's personality and he never managed either to banish one of them or fix the two together.

True to my own one-sided nature, I'm not going to give you a complete portrait of Lawrence here. I'm not interested in the war hero, but the war hero's shadow, or more accurately a series of shadows, a whole forest full of them. The first is my dad, as already mentioned, himself shadowed by his lost name, an independent persona existing now only on cheques and bank cards, called out for in doctor's waiting rooms without response. Lawrence the man died in 1935 after he swerved to avoid two boys on a bicycle and ploughed his motorcycle into a hedge. He hung on to life for a couple of weeks but was never going to recover. In the same year, Jonathan Cape finally published the full-length version of his book, *Seven Pillars of Wisdom*, which had been knocking around in various private and subscribers' editions since 1926. My father's older brother was born just too soon to take on the Lawrence mantle, so my dad represented the first opportunity to honour the fallen hero's name. And yet a few years later the name was withdrawn.

There is a simple explanation. *Seven Pillars of Wisdom* is almost seven hundred pages long, a total of 280,000 words. Thinking of those pillars, Nabokov felt that 'two would have been enough'. He was perhaps not alone. An abridged version, entitled *Revolt in the Desert*, was published by Cape in 1927. This was roughly half the length, cut out most of the poetic description, the philosophical flights of fancy and some of the more personal moments and focused instead on the details of the military campaign: Lawrence stripped of shadow, a wraith. I don't think I'm being patronising when I say that it's unlikely my grandparents had read *Seven Pillars* in 1938, though they may well have taken a look at *Revolt in the Desert*, or at least heard the tales of derring-do it contained. That means they probably didn't know about the episode when Lawrence was briefly captured at Deraa, four hundred and fifty pages into the complete book.

In one of the most famous—and most disputed—sections of *Seven Pillars*, Lawrence claimed that the Turks tortured and raped him before, for some reason, letting him go. This admission alone would have caused a degree of unease in the average 1930s reader, but Lawrence's description of the incident treads close to some form of torture-porn: 'I remembered smiling idly at him, for a delicious warmth, probably sexual, was swelling through me: and then that he flung up his arm and hacked with the full length of his whip into my groin.' Word about this kind of thing gets around, though much more gradually back then than in the age of the internet. My father believes that at some point a few years after his birth someone mentioned to my grandad that Lawrence had been buggered by the Turks. I imagine it happening in the pub, Pop (as he instructed his grandchildren to call him) topping up his pint with a bottle of pale ale, everyone indistinct and grey in the cigarette smoke, Dad sitting on the front step waiting for him to come out, and some smart alec leaning over to take the mickey: *Ain't your boy called Lawrie?* Telling him. And then he was Lawrie no more.

This might seem like a strange place to start with Lawrence, but that episode in Deraa is a possible key to this divided man. Biographers have argued ever since about whether the episode actually happened (why, for instance, having captured and tortured him, would the Turks have let him get away? *How* did he escape?), or happened in any form recognisable from his account. Perhaps the most astute commentary comes from Edward Said: 'I think Lawrence enacted...a ritual—of exposure and being caught...I think we have to simply go with the notion that Lawrence *believed* it to have happened, or believed it necessary to have experienced something like that.'

One thing is certain: Lawrence finished the war burdened with an ineradicable feeling of guilt. He knew from quite early on that the British had no intention of honouring the commitments he was making to an independent kingdom of Arabia. He lived out a lie, and the fact that he believed in the lie himself made very little difference. He was obsessed by the ancient world and also the medieval period. His imagination was populated by knights on horseback, by Malory and the works of Chrétien de Troyes. He wanted to create some sort of exotic other-Albion and he wanted to believe British foreign policy had meshed with his own fantasies. The guilt of simultaneously knowing this wasn't true was either heightened, sublimated or enacted by the writing of the Deraa episode. In 1924 he could tell a friend that the more *Seven Pillars* cost him the better: 'It's part of my atonement for the crime of swindling the Arabs to continue to lose money over my share of the adventure.'

It was in August 1919 that he became *Lawrence of Arabia*. A couple of years earlier, the American journalist Lowell Thomas had been sent out to the Arabian conflict with a cameraman in an attempt to find a good news story which might persuade the US to join the war. There he had come across and been fascinated by the little man in Arab dress who was introduced to him as 'Lawrence, uncrowned King of Arabia'. After the horrors of the trenches, it turned out that Britain needed a good news story, too; a tale involving chivalry, pluck and endurance, played out on camels instead of snowy-white chargers. The same hunger explains why narratives concerning the Royal Flying Corps were so popular (narratives which still had currency over fifty years later when, as I mentioned earlier, I imagined my bicycle as a Sopwith Camel). Using the film shot a few years earlier, Thomas presented a talk about the Arabic campaign at the Royal Opera House in central London. It ran for two months and over a million people came to see it, including the Royal Family, the Prime Minister, Lloyd George, and most of his Cabinet. 'He has invented some silly phantom thing,' Lawrence said, 'a sort of matinee idol in fancy dress…it's so far off the truth that I can go peacefully in its shadow without being seen.'

The conventional wisdom is that Lawrence spent the rest of his life running away from this shadow. In 1923 he enlisted in the Royal Air Force as an Ordinary Airman under the name John Hume Ross. Bizarrely, considering the use to which his own story had been put, he was first interviewed by Captain W. E. Johns, who would go on to write the Biggles books. When the *Daily Express* found out and broke the story, Lawrence was sacked. He joined the Tank Corps instead, this time under the name T. E. Shaw, but he was discovered here too. Eventually he would be readmitted to the air force, always as a private, always hinting that he was serving some kind of penance. But what if *Lawrence of Arabia* wasn't the shadow he was running away from at all? What if he was telling the truth when he claimed that this shadow was useful to him? It's worth noting that, among the million people going to Lowell Thomas's show, Lawrence himself went *at least five times*. He seemed to find it amusing. What if the shadow he was really running from was living in that hut behind the petrol pump in a yard in Epping Forest?

In 1969 *The Times* ran an exclusive story about Lawrence's time in the Tank Corps. A fellow soldier, John Bruce, claimed that Lawrence had paid to be beaten by him with a birch switch or a whip at regular intervals. Apparently the hero of the Arabian campaign had told Bruce that he had stolen money from his uncle, who was now funding this private punishment. In a BBC documentary in the 1980s, his younger brother, A.W., was keen to emphasise that Lawrence didn't do it to get himself off, that the beatings were anti-sexual in nature. 'He hated the thought of sex,' his brother said. 'He'd read any amount of medieval literature about characters—some of them saints, some of them not, some ordinary people—who'd quelled the sexual longings by beating. And that's what he did.' Indeed, right at the opening of *Seven Pillars*, Lawrence talks about soldiers who, out in the desert, 'began indifferently to slake one another's few needs in their own clean bodies—a cold convenience that...seemed sexless and even pure'. In contrast, 'several, thirsting to punish appetites they could not wholly prevent, took a savage pride in degrading the body, and offered themselves fiercely in any habit which promised physical pain or filth'. Lawrence had been degrading his body with pain since adolescence, with gargantuan walking and cycling tours, with fasting and lack of sleep, physical attacks on his physicality. His fear or hatred of sex pre-dated anything that happened to him at Deraa. It went right back to childhood. And all that waited for him out in the wood was happiness, fulfilment, love.

Vyvyan Richards got to know Thomas Lawrence at Jesus College, Oxford. Two years ahead of Thomas, Vyvyan heard mutterings about an interesting, unconventional young fresher and invited him to tea: 'a slight figure with longish flaxen hair and an unforgettable grin'. If Vyvyan wasn't exactly *openly* homosexual, he was about as open as it was prudent to be sixty years before its legalisation and only just over a decade after Oscar Wilde's release from prison. 'It was love at first sight,' Vyvyan wrote. 'He received my affection…eventually my total subservience as if it were his due.' What he didn't receive was any physical expression of Vyvyan's adoration. The relationship remained platonic, any hint of sexuality sublimated into activity, constant activity. They cycled out to look at churches together, discussed the poetry of Christina Rossetti and, developing a keen interest in William Morris, went on an excursion to examine one of his Kelmscott editions of Chaucer. They fantasised about starting their own fine printing press in a similar style to Morris, metal forcing ink onto the surface of thick paper. In their voluptuous, bibliophilic fantasy, they would use only monotype, the 'most exquisitely complicated and precise of all machines'. The pair imagined that together they would print the book Thomas was planning, a study of the key cities of the Middle East called *The Seven Pillars of Wisdom*. They would've been inseparable if only they had ever come together.

When Vyvyan graduated, he got a job teaching history at Bancroft's, a private school in Woodford tucked between Chingford and Buckhurst Hill, right on the edge of Epping Forest, about three miles north-north-east of my home in Walthamstow. The young teacher rented some land and built a hut on Pole Hill, less than two miles from the school, which he used as a retreat. Thomas came to visit and, as early as 1911, the two of them began discussing the idea that they would house their press there. The younger man even went as far as asking his father for a loan to build on the land, but was turned down. Lawrence's commitment to the idea would wax and wane over the next few years. While he was working on an archaeological dig at Carchemish in Syria (and having some kind of relationship with a young Arab nicknamed Dahoum) the idea became less attractive. But shortly after the war ended, he turned up at Bancroft's, was told Richards was out in the forest with his scouts and went to join him. Vyvyan's original hut had burnt down and in its place he built Cloisters, apparently excavating a diving pool as well.

By September of 1919—a month after the Lowell Thomas show began—Lawrence was writing to Richards to tell him that he had bought the land at Pole Hill from under him. He was likely to have another three hundred pounds in six weeks, he said, 'and we must then interview builders'. At Lawrence's instigation, the architect Herbert Baker came out to see the site and make suggestions. 'His dream of the perfect hall, to crown our perfect hill,' wrote Richards, 'was often in our casual talk.' The aim of the design would be to soothe Lawrence's 'worn nerves'. No mention was made of Richards's nerves, even though he had fought in France. 'All was to be for deep quietness in that hidden court: so that the whispering rustle about us of the forest leaves, stirred by moving airs, would sound like the waters of a sea.'

That was the moment, immediately after the war, exhausted and horrified by what he had seen and done, by the death of two of his brothers and by his own complicity in the betrayal of the Arabs, that Lawrence seems seriously to have considered retreating to the hut on the hill and becoming a gentleman artisan. Then he ran again, and kept on running, not from the world but from himself—air force, army, air force, the land sold, no refuge to be found at his next retreat, Clouds Hill, walking out, climbing onto his motorbike and riding, riding, hour after hour, always looking for a turn-off, a narrow side road to a different world, for any sort of escape:

'I imagine leaves must feel like this after they have fallen from their tree.' But no escape was possible until he stopped and looked behind him.

And I picture that other shadow Lawrence—the one he thought he was running from—turning his back on the figure fleeing him, sighing and shrugging, sorry to have been left, sorry for the leaver, lonely there in the early evening. Walking, regretful, back up the hill to the hut, knocking and letting himself in. Stepping over to where his friend sits by the stove looking up at him, bending down, kissing him. The pair of them staying there, inside my head, curled together in one bed, both *meum* and *tuum*, producing the most beautiful books imaginable, the blacks as dark as the darkest sky, the reds the closest they come again to blood. Where they are inside and outside all at once, children and men, knights and outlaws, protected and free.

I decided to pretend that my lurches into sensibility were the product of my quest rather than the zigzags of an inconsistent personality. In particular, I fixated on how different trees summoned from me vastly different moods. The hornbeam made me melancholy in a small, accepting kind of way. It is the most *plastic* tree in the woods, its trunk twisting on itself, strangely self-effacing. A few years ago, when John Kerry ran for the presidency of the United States, I read an article in which he was said to resemble a tree in a petrified forest, a description so apt, so redolent of his cartoon arboreality, that I have never again been able to see him without thinking of it. I found now, in a strange reversal, that when I saw a hornbeam, my first thought was of John Kerry and, perhaps due to his heavy-browed, hangdog demeanour, this made me sweetly sad. On

the other hand, when my eyes jangled with the sudden apprehension of a spray of white birch trunks, saliva would flood
my mouth and a rare exhilaration would grab me, as if I were
ready to go hunting, caught in some ancient Germanic blood-
lust. Beeches reassured me, even while looking like bizarre
psychosexual hallucinations. Oaks I admired as if they were
technically ingenious architecture, without romance. Ash and
rowan drew my eye upward to the spray of their leaves, so that
I always associated them with light. Holly continued to annoy
me, to block my path, to pull me back, to niggle.

Unable to find a hut of my own to hide in, I opted to use
the trees, as if I could teach myself insouciance by trying, tran-
scendence by climbing. If you go to the dogging car park in the
north-eastern corner of the land round Hollow Ponds, cross
over the road and continue into Gilbert's Slade—swinging
round Forest School, another of the private institutions which
dot the London/Essex border—you find a whole commune of
hornbeams crouching in the woods about a mile north-east of
my home, snuggling up inside the North Circular orbital road.
They've been heavily pollarded in the past, so that they tend
to split and twist into agonised yet oddly English postures, as
if they've been caught going through the kitchen cupboards
at a neighbour's party. According to Pevsner's *Buildings of
England*, 'The pride of [Epping] Forest is its hornbeams. There
is no larger forest of hornbeams in England, nor perhaps in
the world.' I thought I should at least try to get up one, for the
sake of local pride if nothing else, and these specimens were
sturdier and—due to the pollarding—looked like they might
be easier to ascend.

There were, though, problems with the plan. As an over-
flow from Hollow Ponds and with huge tracts of housing on

each side, plus a busy road right up close and that massive private school—all pristine playing fields and state-of-the-art gym facilities—this isn't the quietest part of the woods. And I was embarrassed to be caught up a tree, partly because it seemed childish and irresponsible in some way and partly because I was so bad at it. If you're going to appear childish and irresponsible you at least want to do it with some style, to look like a daredevil swaying high up above the canopy on impossibly thin stems, not hanging from the lowest branch, breathing heavily, caught in a losing battle with gravity and age. You want, in short, to be Millican Dalton instead of William Ashon.

Millican Dalton was born in Cumberland in 1867 but moved with his mother and brothers to north-east London after his father's death in 1880. The family eventually ended up in a large house on the corner of two roads in Highams Park. These days Highams Park, which sits between Walthamstow and Chingford, is where London definitively becomes suburban—although I say that as someone who lives in Walthamstow and it's very possible that Highams Parkers feel the same about Chingford. People in Highams Park look down on people in Walthamstow and people in Walthamstow reciprocate in kind. Such is the politics of neighbourhood. Millican, the eldest son, took the top bedroom of the new house, which had a sort of tower or turret jutting from the roof (sadly the house has been demolished). Getting hold of some climbing rope, he began entering and leaving the property via this high bedroom window, then, having mastered the technique, taught his brothers and other local children to do the same. Imagine a big suburban house with a string of urchins moving up and down its front, something from a treasured children's book. From here it was a short journey for Millican and his brother Henry to the woods, where they began tree-climbing in earnest—or as they called it, *boling*—then camping, first for a night, then whole weekends, then longer during school holidays.

Millican's interest in camping continued to grow and develop. Influenced by Thomas Hiram Holding, a tailor considered to be the founder of modern camping, who would go on to write *The Camper's Handbook* in 1908, Dalton became obsessed by lightweight camping equipment which he could carry using a pushbike. When he left school he went to work as a clerk in a fire insurance company in the City, but what was central to his life was getting outside, exploring, climbing things, mucking about. In 1904, like Melville's Bartleby saying 'I would prefer not to', Dalton threw in his job. Except it wasn't a refusal so much as an escape: 'I gave up my job in the commercial world and set out to seek romance and freedom.' He gathered up his camping gear, loaded it onto his bike and headed north to the Lake District, where he based himself in Borrowdale and offered his services as a mountain guide and general-purpose *Professor of Adventure*. Adventure took all kinds of forms, from camping to walking the hills, to rock climbing to rafting on Derwentwater.

Dalton was a genuine nonconformist. While his mission —'to seek romance and freedom'—could belong to Lawrence, his willingness and ability to step outside of the conventions of his day liberated him. Like Lawrence he was an admirer of George Bernard Shaw, but unlike Lawrence he followed through the consequences of that stance, embracing both vegetarianism and pacifism. He was one of the first guides in the Lakes to take women camping, walking and climbing and, once she'd got to grips with it, could often be found following his close friend Mabel Barker up a rock face. He dressed in home-made or modified clothes—corduroy shorts and a shapeless hat set off by a feather. For many, many years he lived in a cave on the eastern slopes of Castle Crag. If you go there now you can still see where he carved DON'T!! WASTE WORRDS! jump to conclusions! on an outer wall after a particularly heated argument with a Scottish friend (hence the double 'r'). Down by the river Derwent are the hazels from which he collected nuts for protein.

But the Lakes (where he became known as the Caveman of Borrowdale) are really only half of his story. A migratory animal, he would return south at the end of the climbing season and shelter in or near Epping Forest, using his mother's house as a business address for the next season's bookings but camping up in the woods. Despite the 1878 Act empowering the Corporation to move on 'gipsies, hawkers, beggars, rogues and vagabonds', Dalton's presence was not only tolerated but treated kindly. They even allowed him to build a hut with a wood burner, where he housed an open library of climbing and outdoors books, would make tea for the Forest Keepers or help kids improve their tree-climbing. He would skate on Baldwins Pond when it froze over and use handmade skis when there was snow, popping over to his mother's to check on his correspondence. Most important of all, he continued his own boling exploits, as if this act of stepping off the ground was what liberated him from the norms of society.

In Italo Calvino's *The Baron in the Trees*—a novel about a young noble in eighteenth-century Italy who climbs up into the trees at the age of twelve and resolves never to come down—the baron quickly sets out his liberatory philosophy:

'it's the ground that's yours', he tells the young neighbour who will later become his lover, 'and if I put a foot on it I would be trespassing. But up here, I can go wherever I like.' You can be sure Dalton went wherever he liked too, my attempts to follow him hopelessly inadequate, a scramble into the lowest branches. The man was a subversive without trying to be, just by the nature of how he chose to live his life. During the First World War he was arrested as a suspected German spy, a case that went all the way to court before being thrown out, hairy pacifists wandering the woods of Essex of course being of intrinsic value to the Teutonic war machine.

There is a problem, though. The Caveman of Borrowdale's liberation was at best partial. In much the same way that people talked about Lawrence, Mabel Barker described Dalton as essentially sexless. I'm (almost) sure that there are people who, for whatever reason, have no interest in sex. I'm just not sure that Dalton was one of them. There's a hint that some sort of breakdown, a mental crisis, had precipitated his exit from the insurance company and his story mirrors Lawrence's too closely in other aspects not to see some kind of sexual equivalence as well. This leaves us stuck in the branches of a contradiction. If I'm going to say that Lawrence's life was one long retreat from his own sexuality, a sexuality which he should have turned and faced up in the hut on Pole Hill, then how can I present Dalton, in his hut a few miles away, as a romantic figure, a baron in the trees, instead of another study in flight and repression?

The poet George Barker was born in Loughton in 1913. I think of him in a trinity with Dalton and Lawrence much like that of Clare, Wally Hope and Will Kemp. Where the other two opt for abstinence, Barker relies on promiscuity, equally compulsive. With fifteen children and multiple affairs, he had enough sex for himself, Dalton and Lawrence put together. In *The True Confession of George Barker*, a poem which seems to have drawn much of its rhythm and imagery wholesale from Clare's version of Byron's *Don Juan*, he states 'The pleasure of money is unending / But sex satisfied is sex dead. / I tested to see if sex died / But, all my effort notwithstanding, / Have never found it satisfied'.

You must have guessed by now that we're never going to get out of this forest of text without meeting Freud. In 1912 he published a paper entitled 'On the Universal Tendency to Debasement in the Sphere of Love'. In this work, Freud takes the psychoanalytic explanation for male impotence— that it is the result of a disconnection between the *affectionate* and *sensual* currents resulting in an inability to have sexual contact with the object of one's affection—and applies it to society as a whole.

The original object of affection (for a man, presumably) is always the mother or sister, Freud claims, and this affection carries an erotic charge. As this is a forbidden or taboo connection within civilised society, it is very hard for *anyone* correctly to connect this affectionate current with the sensual current that rises up at puberty, particularly as most civilised societies (the term is Freud's) also restrict the pubescent child's sexual freedom of expression, making them wait for any fulfilment. Freud's conclusion is that impotence is merely an extreme manifestation of a generalised problem which causes men to degrade (or choose degraded) sexual partners and women only to enjoy *normal sensation* within a secret love affair. 'It is quite impossible to adjust the claims of the sexual instinct to the demands of civilization…renunciation and suffering, as well as the danger of extinction in the remotest future, cannot be avoided by the human race.'

But, Freud goes on to say,

> the very incapacity of the sexual instinct to yield complete satisfaction…becomes the source, however, of the noblest cultural achievements which are brought into being by ever more extensive sublimation of its instinctual components. For what motive would men have for putting sexual instinctual forces to other uses if, by any distribution of those forces, they could obtain fully satisfying pleasure? They would never abandon that pleasure and they would never make any further progress.

Perhaps this explains Freud's use of 'sphere' in that title, despite his belief in 'progress'. The continuing development of the culture of civilisation is driven by the very dissatisfaction which its development creates. We're stuck in a circle, wandering round and round, denied satisfaction. Our sexual frustration is why we write books, make art, design jet engines and, possibly, bomb each other.

Back at school, one of my fellow pupils invited Danbert Nobacon, lead singer of second-wave anarcho-punk band Chumbawamba, to perform songs from his acoustic solo album, *The Unfairy Tale*, in one of our meeting rooms. Midway through the evening, between two songs, Danbert got out his knob in order to discuss how little it looked like a microphone, or for that matter, a missile. At the time I was preoccupied by not examining Danbert's penis but now I see it as a commentary on this theory of Freud's, that society will only continue to develop, to improve itself, to the extent that we're none of us getting properly laid, that the whole edifice of our achievement is just a rather involved dick joke: sexual repression as the motor of human 'progress'. Head back to Dr Matthew Allen and his belief in avoiding 'even the appearance of unnecessary restraint', in always treating his patients as 'rational beings' and you see the same separation between rationality/morality and the *animal urges*, though recast to the temper of his times. Both show us a kind of self-fencing or auto-policing of the psyche which may begin with sex but which spreads out across our lives—or, just as plausibly, reaches into sex from some other part of our lives.

Imagine if, counting the huge cost of enforcing and policing Enclosure, we had all decided that it would be easier and less painful if we did it ourselves, if we learnt to respect private property, private *land*, as if it had been given to its owners by God, not taken by force. Why then, having set up, unregulated, our own internal policeman—a tiny, insistent PC Connor of cowardice and restraint—would that brutal, angry little bobby limit his authority merely to issues involving property? England has not had a full and complete record of land ownership since the *Return of Owners Land* in 1873. Only four years later, Annie Besant and Charles Bradlaugh were tried and convicted of publishing an 'obscene libel' for printing Charles Knowlton's *The Fruits of Philosophy*, a book which promoted the use of birth control. If the life of John Clare shows us anything, it's that our attitudes towards land and the body are intertwined, just, indeed, as our bodies are intertwined with the physical space they inhabit. Is it possible that, in limiting and repressing our relation to the land, we have limited and repressed our relation to our own bodies too?

When Freud was trying to explain the slow beginning of *The Interpretation of Dreams* to his associate Wilhelm Fliess, he described the book in the following terms: 'First the dark forest of the authors (who do not see the trees), without prospects, rich in false paths. Then a hidden gorge, through which I lead the reader…and then suddenly the summit, and the view and the inquiry, "Which way would you like to go?".' Why, with this view laid out in front of them, this landscape made navigable, understandable, *controllable*, why, having been led to enlightenment, would anyone ever want to go back to the forest again?

Back in the forest I was stuck up the hornbeam. It's a universal truth of tree-climbing that, however hard it is getting into a tree, getting out is harder. You can pull yourself higher by a stumpy, suction-cup handhold but try lowering yourself back the same way and your face will meet bark. Pollarded trees in particular have a tendency to bulge out, to *crown*, so that when you come to look back at the floor you can't even see the trunk to navigate down it. The hornbeam I'd chosen was cloven—or unconjoined, depending on how you looked at it—at the point where two trunks had previously come together, so that it looked like the x-ray of a mutant wrist: radius, ulna, pollard-lump palm then a craze of fingers. I couldn't sit once I was up there but it was comfortable to stand and lean back against one of the thick branches twisting skyward. It was a very deep grey, subtly banded like the leg of a huge cat, and I felt that I should take a moment for quiet contemplation, a meditative pause. I couldn't. I was already worrying about how I'd get out.

There are no low branches on the Cloven Tree, none at all, none left from previous loppings, none jutting out at ninety degrees to the trunk. They all continue in roughly the same direction, heading away from the ground towards light. There is nothing to swing from, nothing to allow you to lower your feet—toes pointed, searching—back to earth. I knew it before I went up but allowed myself to believe I would figure something out. Now I'd arrived it didn't look so smart. The private school would be chucking out soon and no one wants to be caught hiding up a tree overlooking a path near a school. After assessing all options and trying a couple of scrabbled, desperate retreats that had something beetlish about them—as if I were already stuck on my back with my legs scratching frantically at the air—I decided I would have to jump. It was probably only two and half metres, though it looked further inside my head. I went before I thought about it too much, a clumsy slither from the branches before I fell momentarily, a lurch through space ending almost before it had begun with my misjudged, crumbling contact with the ground. I tried to convert my downward momentum into a panicked run but tripped, fell further, ended up sprawled, face down in wet leaves. I stayed there for a minute, first to make sure I hadn't hurt myself, then longer, to remind my body both where it came from and where it was going back.

CHAPTER 11

Ongar Park | Comical Corner

I t was winter when I went to visit Penny again. It was a cold, bright day and I had intended to walk to Dial House, a plod of some fourteen miles or so. I used the excuse of half-term but I was underprepared for our meeting and spent the morning and most of the afternoon trying to sort myself out before I drove over, grumbling up the M11 between rush hour queues at every junction. I was meant to be staying the night but I backed out of this, too, citing the same school holiday and unspecified busyness, which wasn't exactly a lie but not exactly true either. A couple of weeks before I'd had a dream in which Penny had climbed into bed with me and leant in for a kiss. I had cradled his head while avoiding his mouth, so that his stubbled cheek had rested on mine and we had lain there until he slept. The dream had stayed with me, ineradicable, its atmosphere polluting a reality it bore no relation to, and I'd be lying if I claimed it hadn't played some part in my last-minute decision not to stay. It's hard to say what my subconscious had cast Penny as—father? son? lover? abuser?—but it may have been as Death. Whatever its message, it made me drag

269

my feet. I arrived just as the sun had sunk beneath the horizon and the sky almost vibrated in its rush to deep blue up behind the black of the house's eaves.

I'd come back to Dial House to talk to Penny about Enclosure, about escape from Enclosure and resistance to Enclosure. It kept cropping up, a tall fence, whichever direction I tried to walk through the forest. If the English land-grab of the eighteenth and nineteenth centuries was the prototype, it was one foreshadowed right back in William I's annexing of huge swathes of country for his hunting land, then repeated first in the Clearances in Scotland and in more distant colonial expansions. It is, moreover, a process still enacted all over the developing world to this day, from the jungles of Chiapas to the swamps of the Niger Delta, from Wukan to Sudan to Siberia and the Artic Circle.

Marxists drawing on Karl's notion of primitive accumulation have come to see this expansionary privatisation as an ongoing process, the still-gurgling wellspring of capitalism, and it stands at the very heart of debates on globalisation. The ecological movement has embraced the notion of Enclosure as a way to understand corporate incursion on the environment. In addition, the metaphorical power of Enclosure as a way to describe the privatisation of public space impacts on everything from debates about shopping malls to the internet. We've seen that in Ubi Dwyer's explanation of the Windsor Free Festival in the early 1970s, and Paul Morozzo's justification for Reclaim the Streets in the early 1990s. The very phrase *Creative Commons*—used to describe a new approach to copyright—has its roots firmly in the soil of the English rural past. It is a crisis which never seems to resolve, a split which we have so far come up with no way to heal.

Dial House sits on land that was once part of the estate of Ongar Park Hall, originally a small section of the Stanford Rivers estates but parcelled off as part of a settlement between three brothers sometime in the mid-fourteenth century. At least I think it was. I was distracted again and again during my research by the fact that William de Monceaux at nearby Nash Hall was awarded a marshalship in 1212 which made him responsible for looking after the prostitutes at the king's court and chopping up condemned criminals. Ongar Park Hall couldn't match that, offering only a dribbling succession of owners who came and went, the property changing hands on the indebted death of one prodigal son or another with all the inevitability of a physical law. From the late seventeenth century it was leased out, for ten or twenty years at a time, to a succession of businessmen with City addresses, reflecting the new realities of capitalism. Local legend states that the hill above it was used as a beacon or watch post during the Napoleonic invasion scares of the early nineteenth century (the same scares which saw John Clare sign up briefly to a militia, more for the wage and food than from any urge to fight), but it wasn't until the 1880s that the North Weald Redoubt was built. It was, in effect, a weapons bank, where local militias would gather to be armed in the event of an invasion. In 1920, the Marconi company moved in next door and built a radio transmission station sending Morse Code messages to Paris. After the Second World War, the station was nationalised and made part of the General Post Office. In 1980, British Telecom was floated off from the GPO and in November 1984, in one of the Thatcher government's landmark privatisations—and only four months after Crass had split up—it was sold. The seven-hundred-acre parcel of land round the radio station, including Ongar Park Hall, a farm by this stage, was now in private hands. Throw in the technological developments which meant that the radio facility was on the brink of obsolescence and you can see where we're going.

By 1988 BT were beginning to look into the 'development potential' of the land and in 1990 they unveiled their proposals. As Penny remembers it, 'they'd planned a massive hotel at the top of the hill, then luxury houses. Dial House and the farmhouse were going to be the centre of a bijou *hamlet* with shops and heritage attractions and all that stuff.' He was not best pleased. Nor, it turned out, were the residents of North Weald Bassett. Let's be honest, North Weald Bassett isn't the prettiest village in the world but it has a large common running right up behind it—that green hill with the pleasant view—and, due to the easy-going attitude of the tenant farmer at Ongar Park Hall, the common extended, in effect, right across his land. Penny and Gee decided that they would fight the development and found no shortage of assistance down in the village.

'We formed an action group, a village action group,' Penny recalls. The group was drawn from across the social spectrum, the kind of ragtag mob of working-class East Enders and local toffs you'd expect in a village suspended between city and country. One of the immediate advantages over other such groups, though, was the presence of sometime punk troublemakers in their midst: 'We used an awful lot of the skills we'd learned in the PR department of Crass!' A monthly newsletter was delivered to every house in the village. The local chemist gave up a spare shop he had sitting empty and this became an information centre and café, a meeting place for the resistance. Without Dial House and the lifestyle of its residents, without the time and space to deal with the nitty-gritty of the case, Penny thinks, they never could've won. Dial House had been a listed building since 1984, but Gee managed to get the farmhouse and, crucially, all the outbuildings of the farmhouse listed, making any development of that area nigh on impossible. She and Penny worked closely with the lawyer who had taken on their case. They drafted in experts such as the Cambridge woodland historian Oliver Rackham, and prepared legal papers. 'We put together a brilliant case, fought it, fought the appeal. And won. Which was a pretty spectacular success, actually. And really, in terms of public statements, there's nothing I've done in my life which is more successful. In a way, it's a better story to be told than Stonehenge or Crass or the house as an experimental thing. To have had such a spectacular victory against one of the biggest corporations is in most people's minds unthinkable. But we did it and we didn't do it on our own. We did it as a community.'

That was 1992, and the residents of Dial House assumed it was all over. But British Telecom decided to cut their losses and sell the land and buildings. In 1994 it was bought by a company called Peer Group PLC operating on behalf of London & Continental (Holdings) Ltd, just two of a knot of companies operated out of an office in Southwark. Peer Group, Penny alleges, went to war with him. 'We could use British Telecom against themselves—because corporations don't like being unpopular, they've got a very strong PR department who don't want to be embarrassed. But some crap holding company who were basically hoods? They were just vicious. It started getting unpleasant.'

The company calculated, Penny believes, that the best way to further their long-term developmentt plans was to get the

273

residents out of Dial House, and that meant Penny—or rather Jeremy John Ratter. They pursued this policy, he says, with 'conventional landlord tactics', telling him, for instance, that if he did anything to the exterior of the building he was in breach of his tenancy agreement and then leaving the kitchen roof leaking for a whole winter. Then, in January 2000, Penny had to attend Cambridge County Court for eviction proceedings. 'It was,' Penny says, 'more like a criminal trial than a housing issue. They were trying to claim that I hadn't got any right to be here—through character assassination, really. But they hadn't done their homework. They hadn't even spent a couple of days on the internet or they'd've had a much better case!'

As it turned out, the magistrate ruling on the case was himself an architect with a fondness for Charleston and the Bloomsbury set, 'and immediately he saw this place you could tell he was in love with it'. He ruled that Penny could stay, partly on the precedent-setting grounds that, as a professional writer, he needed the extensive gardens for 'contemplation'. Penny can do contemplative when and if necessary. Obviously the magistrate hadn't been made aware of the defendant's other comment at this time: 'If I'm going to close down thirty years' commitment, it's not going to be because some yuppie bastard has decided to make my house into a gold mine. What they'll get is a burned down ruin.'

Eventually, in March 2005, the company put the house up for auction, sold with Penny as sitting tenant. Concerned that Peer Group would do everything they could to prevent them buying the building, Gee and Penny instructed someone else to bid on their behalf. They agreed the night before to go up to £160,000. 'This was all whimsy, really,' he says. Despite benefit gigs and appeals, they didn't have that kind of money. But their bidder won the auction, going right up to that arbitrary limit. Fifteen years after BT unveiled their proposals, Dial House and its inhabitants were safe. But crisis grips tight, like Jacob and the angel.

'Weirdly,' Penny Rimbaud recalls, 'I then suffered the worst depression of my life. Because I never wanted to own and I didn't like owning. I knew it was the sensible thing to do but at the same time it was something that went against my whole ideology, really. Because before then it was everyone's and suddenly it became mine and Gee's.' This depression was compounded when the two people who originally said they would back the purchase and pay for the house both pulled out. 'So I did the worst thing you can possibly do. I went to my lover, who was Eve [Libertine], who had some money through inheritance, and my sister, who had money, and my uncle. And of course there's nothing worse than borrowing from lovers and family. And for me it created a problem in all of those relationships. I never felt I could be myself again.' Sometime after this Penny and Eve split up and I can't help feeling all this must have played its part.

Only now, another ten years on, are all those debts repaid. And, when the last of the nominal interest payments has finally been dealt with, Gee and Penny are planning to place Dial House in trust. 'Ironically it means you give it to the state!' he told me, laughing. 'But you give it to the state with an agreed purpose.' The pair of them have settled on describing Dial House as a centre for the radical arts and horticulture. 'Given our history,' he explains, 'a statement for the radical arts alone might be too much for the state to agree to, but adding the permaculture thing should be enough.' I can see it now. Dial House will end up, at some future point perhaps fifty years forward, as a shrine to gardening and a lost conception of 'truth and beauty' and little old ladies and gentlemen who used to be punks back when they were young will drink Earl Grey out on its lawns. Their children will wait out in the car park, thinking but not saying that it's all a little quaint and silly while their parents potter, shaking, through the gift shop. But those children will be wrong. I hope that one of the paths I've followed through these woods—I have no idea which—will suggest why.

Since I'd last seen him, Penny had suffered what he described to me in an email as 'a fairly massive heart attack'. I may have been projecting but he seemed a little older, slightly more unkempt, the way his long hair hung from under his black woolly hat making him look like a cardinal who'd been at the communion wine. Except Penny doesn't drink any more and, after the heart attack, had rather gone off his roll-ups, opting instead for enthusiastic, almost incessant vaping, freely admitting he had no idea how much nicotine he was taking on board. When I arrived Gee was out teaching a t'ai chi class to the old(er) folk of North Weald, Dominic was cooking again, though not fresh bread this time, and Gordon, who I met when I first came here—and who may well have been responsible for introducing Penny to e-cigarettes—was polishing an impressive set of biker's gauntlets and engaging me in conversation about Great Road Snarl-Ups of the M25 and M11. I didn't ask Penny about the heart attack, not even when we got out to his shed and he was sitting down at his desk and I had my recorder running. I suppose I was embarrassed to, or didn't know how to raise it, or didn't want to seem glib. But we got there, anyway, because it illustrated, to Penny at least, what it was we were actually there to talk about.

He was in Rochdale when it happened, waiting on the platform for a train to Manchester, on the way to do a Q&A at this festival or that. 'There was no forewarning,' he said quietly, rolling his e-cigarette round in his hands, his thumbs the thick stubs of someone who had, at some point in his life, used them for manual labour as well as drumming. 'It just kind of *whammed* me. The power of it was actually quite exquisite. I've never been in a situation where I *couldn't do anything*. It was like this massive tsunami of experience pouring over me and pushing me down and down and the more I tried to rise through it the less I succeeded. And only by going with it was I able to…go with it!' In the ambulance he dipped in and out of consciousness, meditating in an attempt to ride the pain. In doing so he had a '*profound* experience of falling into that place where I no longer existed, yet actually everything was just the same'. He paused for a moment, trying to get it right. 'We're so silly about our ideas of survival. Because actually what we're talking about is ego and consciousness survival. We're not able to trust that deeper thing which is actually the very thing which allows us to survive.' He was grasping for something, hopping round it, and my interruptions—which pepper the tape of our conversation, well-meaning additions and awkward attempts to fill any silence—weren't helping.

'It's confusing and sometimes disturbing for me how much baggage I lost with that heart attack. I feel like the train I missed on Rochdale station carried off a person who I knew and left me empty of a large amount of that person. And I'm very grateful. I'm very pleased I had a heart attack for that simple reason. Because it helped me rid myself of a lot of conceits.

And that's that Cartesian thing—I thought I needed all that baggage. And I'm sure I've still got a whole load of baggage. I used to believe that idea that universality was only achievable in what I called the Romeo/Juliet syndrome—y'know, only in death can we become as one. But I now know that we *are* as one. And if we want to imagine otherwise, that's our problem. But it is our problem. And if we're going to do that we have to accept all the problems that creates, i.e. the absurd violence and abuse and the absurd hilarity and overexaggerated entertainment and joy—or what's called joy, which is no joy at all, that entertainment idea. But the extraordinary thing in my mind is that we justify our own Cartesianism or we defend it or we hang on to it, yet actually cannot respect other people's Cartesianism. Because you can't!' And this, right here, was my reason for being in the room.

I t had slowly become apparent to me that I needed to define Enclosure more broadly than a few fields and fences. Why just stick to the physical, material world, after all? Wouldn't that in itself represent a kind of Enclosure? Thinking back to Ken Campbell, I felt I needed to define Enclosure as *any* restriction of access to space—physical, intellectual, emotional, psychic—brought about for private gain. Freud might trace back our core dissatisfactions to bad sex, but the *civilisation* generated and regenerated by dissatisfaction is capitalism and something systemic stops us from seeing—or even admitting—the possibility of any alternative. Take our attitude to ownership of land as an example, in particular our unquestioning acceptance of the status quo. Outside of Scotland, there has been no talk of land reform in the United Kingdom in the last fifty years. Yet it's arguable that many of the economic problems which afflict us flow from it—from the high cost of housing to low growth. It's even partially to blame for the idea that the UK is overcrowded. Advertising, as another example, is a form of Enclosure, shutting off and restricting our range of responses in order to persuade us to consume in certain ways, to aspire to be thin and have good teeth, to own an iPhone or cycle in Lycra. Culture more generally plays much the same game, even when it seems to engage with the problems of the world around it, leading us towards despair and exhaustion, embracing entropy. It's not just the land which is increasingly controlled by private corporations—from shopping centres to garden bridges across the Thames. Our imaginations have also been occupied. Our common language was first centralised and systematised and is now being sold off or used to reinforce the status quo. Politics operates within a thin strip of *realism* and *fiscal responsibility*. A bank which has lost three and a half billion pounds in the last twelve months can pay its senior staff five hundred million in *performance bonuses*. Governments can monitor all our electronic activity to guarantee our freedom. And these examples operate near to the surface, easily spottable. What about the deep grammar of assumptions that limit even our ability to see a problem, let alone see it in a different way? This is not an argument for *false consciousness*, the old Marxist saw. It's an argument about fenced consciousness.

The American social critic Curtis White writes of the Middle Mind, a kind of Enclosure of the imagination whose very purpose is to maintain our social relations as they stand *without us even noticing*. He describes us as an 'administered population' who conspire in our own administration, in the murder of our imagination. 'We are told, of course, that we already have all of the justice, freedom, and creativity we have any right to expect in our Western, neo-liberal, free-market democracies,' he writes. 'We are told this with the sort of intensity that reveals the fear of the consequences if this claim were ever shown to be false. The truth is that this claim is the Big Lie. It's like life in an alcoholic family, where all the family members are daily obliged to reaffirm that everything is whole and right and normal...Every day our social imagination is asked to defile itself by confirming a lie. And every day most of us run from the horror of confirming a lie by refusing to think.' This is perhaps the third stage of the project of Enclosure. First the land had to be held by force—by incarceration, transportation and execution. Next, we learnt to police ourselves. Finally, we taught ourselves to forget that there was anything to police.

At around the time I returned to visit Penny, I also decided I needed to spend a night out in the forest. I thought it would provide a climax to my journey, a sense of destination or a therapeutic release of unspecified nature, some form of magic to glue together the speculation. My first idea was to go to stay on the campsite that Newham Council owns down in Debden—basically because it was safe and easy. I would pick an out-of-season night when no one else was there, have a fire, be utterly alone on the edge of the woods, get up at dawn and walk through the trees, maybe follow Epstein's lead and swim naked in a pond. I knew even as I toyed with the idea that it was a non-starter. I had been to Debden campsite a few times over the last fifteen years and the idea that it could provide some magical conclusion to my journey was laughable. I had sat with other campers and watched England crash out of another major football tournament in a stuffy, overheated room there. A friend's child had nearly been knocked down by a runaway car with a snapped handbrake cable. On our last visit we had been rescued by the AA after losing the key to our own car in the field. If the Debden campsite was a locus of anything, it was bathos. No, I would have to stay right out in the trees.

I had already spent the night—or most of the night—out in the forest once before. Every year at the end of November, the Friends of Epping Forest hold the Rodings Rally. This is an all-night orienteering event in which small teams have to navigate round five or ten checkpoints in the woods, most of which are hidden inside tiny, unlit tents. Just to make it harder, each location has to be picked from three possible sets of coordinates on the basis of a semi-cryptic clue which may or may not refer to the name of the area the checkpoint is located in. As I was to discover, everybody takes this all very seriously. For instance, people from other teams won't talk to you in case they are accused of cheating and end up being given your awful time. This doesn't fully explain why they won't smile at you, either, but I suppose it's dark and maybe they just didn't notice my friendly grin.

I was on the Rodings Rally because my daughter was, at the time, a member of the Woodcraft Folk—traditionally referred to, with very little sense of respect to either party, as *hippy scouts*—an organisation which, inexplicably, flourishes in my area like Japanese knotweed. This particular group is *parent-led*, which means that what you probably already thought was the most middle-class thing ever becomes just that little bit more middle class (which is rather sad or quite reassuring, depending on how you want to look at it, as the Folk were originally set up as a socialist working-class youth movement: 'We are the revolution. With the health that is ours and with the intellect and physique that will be the heritage of those we train we are paving the way for that reorganisation of the economic system which will mark the rebirth of the human race'). All the parents involved are good people and I'm very fond of them, except on the few occasions when I have had to sit in a circle at camp, hold hands with them and sing the Woodcraft Folk song. Anyway, one of these parents suggested we take some of the kids to Rodings Rally and because I thought staggering around the woods in the dark would be preferable to sitting in a hall interacting with people at one of the interminable Thursday-night sessions, I volunteered. I suppose you could say it was a selfish decision.

René Descartes published his *Discourse On Method* in 1637, just sixteen years after Mary Wroth's *Urania* but a long way away in terms of its underpinning assumptions. We're not only talking here about Wroth's Platonic world view, the Renaissance magic and alchemical symbolism which infuse her work. We're also talking about her reliance on *contingency*, on those repeated experimental combinations, on the fact that nothing is ever finished or complete—or, possibly, capable of being finished or completed. Descartes, by contrast, is all about certainty. The whole drive of his project is to find solid foundations on which to build human knowledge. And, when he talks about his method for achieving this, he uses an interesting metaphor.

'In this matter,' Descartes states,

> I would imitate those travellers who, finding themselves
> lost in some forest, should not wander about by heading

first this way and then another, nor still less remain in one place, but should instead always walk as straight as they can in the same direction, and certainly not change their direction for petty reasons, despite the fact that at the beginning it has perhaps been chance alone that determined them to choose that direction; for by this means, if they do not come exactly to the place they wish, they at least will arrive at the end of some part of the forest, where they will probably be better off than in the middle.

The first thing Penny said when I told him about that passage was that it was a silly way to try to get out of a forest. If you just choose a straight line at random, you may end up walking into the deepest, darkest centre of the woods. Yes, you will eventually come out somewhere or other, unless you've been eaten by a bear. But it's hardly the smartest way to go about it. Instead he recommended taking a moment to sit down and listen—'you can somehow *hear* where the centre of the forest is'—then looking at which side the moss is growing on the trees, feeling where the wind is coming from and so on. Penny's point is simple. Any terrain is unique, even inside a metaphor, and you need to react to that unique terrain rather than to a general principle. I might go even further and ask why you wanted to get out of the forest in the first place. Descartes is presumably thinking back to Dante's dark forest and suggesting that rather than following a ghost, one should rely on one's own rationality. It is, however you look at it, whatever you think of it, a classic statement of the fundamentals of Cartesianism—the fetishisation of the perfect geometry of the straight line, wrapped up and disguised as the most self-evident common sense.

It is this straight-line thinking which Descartes uses to reach his famous first principle: *Cogito ergo sum*—I think therefore I am. This is another formulation which Penny takes issue with, finding it to be so *obviously* ludicrous that he almost struggles to explain exactly how, though an understanding of Zen and meditation probably helps. And when I say *understanding*, what I really mean is *immersion in the practice of*. Because when he puts it into language, our language, our common language, with Cartesian assumptions built in—or if not built in then at least bolted on and souping up that language in such a fundamental way that it's hard to uncouple them again—it sounds as if he's pushing at the outer boundaries of sense.

We started simply. 'The "I" is a thought itself,' Penny offered. 'So that's ridiculous—*I the thinker think I am*. I'm really much more interested in what's before the thinker. Because what's before the thinker is the source of the thinker. And the thinker is the source of the thought. So at the very most *I think therefore I am* is the thought in itself. So already it's sort of collapsed itself.' Penny's argument is that the idea of self is indeed itself an *idea*, a thought. It's not only that the phrase 'I think' already implies a self but also that it implies a false impression of self. Because if you follow Zen practice in any serious way, it's the conception of self as a discrete entity which begins to fall away.

What I think Penny and I both agree on is the way in which Descartes's line of argument privileges the 'I', the individual. If your own thoughts are the only thing you can be sure of, then 'I' comes first. If you can't know that anyone else is 'human' in the way you are human, then there is no core reason to treat them as such. It's a simplification, of course, but Descartes could be said to have invented individualism and in doing so to have kick-started Enlightenment thinking. 'You can only be "I the thinker" if you have a point of view or a position from which one is making these static judgements,' Penny said. 'So that's my interest, almost my revulsion, with Cartesian thinking—that it holds one always in this static position as if there is some sort of permanent point of judgement.' We're getting somewhere here, with that mention of revulsion, because Penny's beef with Descartes is less intellectual than it is ethical. 'It's basically a piece of utter nonsense and out of that utter nonsense I think one can point to celebrity culture, to consumerism, particularly to the propaganda of consumerism. You had to create that sort of individual in order to make that individual the target.' The individuated, unitary self is, it turns out, a form of Enclosure too.

When I tried some half-hearted defence of the Enlightenment as the liberator from serfdom and, eventually, from God, Penny replied that it led directly to the Industrial Revolution and the Holocaust. There's something in this, even if it sounds crazy put so boldly. The Swedish author Sven Lindqvist has explored how Western colonialism led to, or created the conditions for, the Holocaust. And colonialism itself can be seen to grow out of an extension of the land-lust which started with English Enclosure (a land-lust that never had anything to do with a need for space so much as for the primitive accumulation which was starter fuel for the capitalist engine). Cartesian rationalism was at the very least used by the gentry to justify Enclosure. As the historians John and Barbara Hammond put it in their classic work on Enclosure *The Village Labourer*, at the time 'the rage for order and symmetry and neat cultivation was universal'. It's also true to say that the self cannot be riven—cannot break apart as it does with

John Clare—until the self exists. It could even be argued that the splitting of self is a response to the narrowness of selfhood placed upon us. Or indeed that it's almost a precondition. After all, the dualism of the mind–body split is inherent in the *proof* of 'I think therefore I am'. By giving existential primacy to thought, you are turning your body into nothing more than the ship of sensations in which your thought sails.

While we are dealing with the self, with identity, we need to talk about names. All of the people we've met in this book have had more than one. Jerry Ratter transformed himself into Penny Rimbaud, just as Phil Russell had become Wally Hope. Richard Turpin tried hiding out as John Palmer. T. E. Lawrence rejoined the army and air force as J. H. Ross and then T. E. Shaw. John Clare attempted to pass off his Elizabethan pastiches as the work of Percy Green, James Gilderoy and Frederic Roberts and, imprisoned at Lippits Hill Lodge, told visitors he was Jack Randall or Lord Byron. Both Margaret Epstein and Mary Wroth had their names changed by marriage, in Margaret's case twice. Wroth, meanwhile, reconfigured herself in her writing again and again, as Pamphilia and Bellamira and Belizia and Lindamira and so on. Mick Roberts, ex-con, became Old Mick, patriarch of the roads protest movement. Millican Dalton became the Professor of Adventure and Caveman of Borrowdale. And while Ken Campbell was always utterly and completely Ken Campbell, he was also Sophie Firebright (and the Spanking Squire).

I nicked the notion of enantiodromia from Campbell and I've used it as a way to characterise a split or sharp contrast, a bifurcation in personality which I've hinted is in some way a typical reaction to crisis and also abetted or exaggerated by contact with the forest. But at this stage we're beginning to slide into something more nuanced, something which can't be contained in only two facets, by mutability and metamorphosis, the idea expressed by Robert Pogue Harrison that 'the law of identity and the principle of non-contradiction go astray in the forests'. Enantiodromia is perhaps best viewed as a simplification of this principle of mutability, an attempt to make it fit with a binary mindset. All of the central characters of this book reinvented and reimagined themselves during their lives, and they used the forest not merely as the site but as the crucible for this alchemical transformation. In doing so, they reimagined the world. Reality, as Olafur Eliasson tries to show in his artworks, is 'something we are constantly negotiating and forming'. Or, as Curtis White has it, 'the reality that we refer to as our daily reality is simply the work of human imagination that has become ossified, codified, and generally naturalized'. By reimagining your self, you open up the possibility of reimagining *everything*.

In 1968 an economics paper was published, 'The Tragedy of the Commons', which purported to show that common ownership could never work because it was in the rational self-interest of each party to a common ownership agreement to exploit that agreement by taking more than their share of the common resource. Although the author, Garrett Hardin, would go on to admit that his argument was flawed, it became one of the most cited academic papers of all time, used as justification for wave after wave of privatisations in the 1980s and 1990s. Because the pasture on which he feeds his animals is given to him for free, Hardin argues, the rational herdsman 'concludes that the only sensible course for him to pursue is to add another animal to his herd. And another;

and another…But this is the conclusion reached by each and every rational herdsman sharing the commons. Therein is the tragedy. Each man is locked into a system that compels him to increase his herd without limit—in a world that is limited. Ruin is the destination to which all men rush.'

Does this remind you of Descartes walking in a straight line chosen at random and followed come what may? The same abstraction from a unique, lived reality, from an actual terrain? The same notion of *rational good sense* resulting in utterly irrational behaviour? It is also underpinned by the same view of humanity and consciousness—of people as discrete thought-packages motivated solely by self-interest, incapable of cooperation or empathy. This is, in fact, the rational actor of classical economics, the very same individual who is used to justify the liberalised market as being, in all cases, the best, most efficient way to distribute scarce resources. But these assumptions can be found everywhere, not just in Margaret Thatcher's claim that 'There is no such thing as society'. There is a thought experiment in modern consciousness theory, pushed hard by the philosopher David Chalmers since the late 1990s, in which we are asked to posit that other human beings could be zombies, that is to say that they act in every particular like us but lack the internal consciousness which makes us human—as if our physical actions are somehow separable from our consciousness and that this 'experiment' doesn't assume exactly what it sets out to prove.

But then again, maybe you can take this line of argument a little too far. Or maybe I just wasn't ready to leap that far away from myself. I grew up living in a materialist world. I have tended to view, unquestioningly, all thought as growing from its social context—to see Descartes as a symptom of early capitalism rather than a cause, I suppose you could say. Penny is familiar with this dichotomy. In an interview he gave recently, which mainly attracted attention for his frank description of wanking off his dying father, he also talked about the tensions that had arisen within Crass between him and bassist Pete Wright. These were, at root, he felt, because 'whereas he was probably essentially a materialist, I am essentially a mystic'. Penny believes—believes he knows—that the world could be different if we all started thinking differently. Which is in some senses a truism and in others the most wide-eyed idealism.

Do not confuse it, though, with the individualism of epistemological idealism. Penny's whole philosophy is based on the notion that the self is unenclosed, that the difference between him and me and you is one of degrees rather than an absolute boundary. That being the case, in talking to me about our selves, he begins to run up against the barrier of a language built on the very assumptions he rejects. When we met before he had told me about his feeling of elation— of liberation, even—on awakening after a serious operation at a traumatic time in his life. I asked him how that compared to the liberation he was feeling now, after his heart attack.

'The interesting thing is,' he started off, 'when I came out of that operation, that sense of extraordinary joy and bliss was really my ego taking control of an experience and using it to maintain itself. The heart attack actually was in some ways a similar experience. Except I don't feel joy in that way any more. I don't. And that's what I mean when I say I get a little bit confused or even a little bit disturbed—at the fact that I *feel absolutely nothing*. It makes no difference. Trying to encompass, trying to catch hold, trying to hold on to. All of those things we try to do. Contain things, own stuff, possess things. All of that stuff is our downfall. It makes us all *leaden*.

'The person who felt so much joy was weighed down with that,' he explained. 'Attachment to joy is just the same as being attached to a Rolls-Royce. It's exactly the same thing. There is nothing but the profound nothingness of the moment, which is all-engulfing. It doesn't laugh and it doesn't cry, it simply is. And I don't claim to have that purity. I certainly have moments where I'm aware...' He stopped, laughing at his own folly, how easy it was to fall into the trap. 'No, I'm *not* aware. You see, that's the wrong word. The moments are when I'm *not* aware of any moment, then the moments actually *be* and I exist entire and within those moments. Their knowledge is so insignificant. It has no hold, it has nothing. Because the knowledge is absolute so therefore has no word.'

The start point for the Rodings Rally is up near High Beach, a couple of hundred metres from the lost Turpin's Cave pub and about the same from the Kings Oak. As my group set off, the parent who had arranged our entry and collected all the details from the organisers muttered something to me about how the first checkpoint was the first on the list. I took this to mean it would be the first set of coordinates of the three on offer and so we started without paying too much attention to the clue, the significance of which had passed me by. My teenage daughter and her two friends quickly linked arms and began chatting among themselves about this scandal or that, making it clear that they weren't interested in map-reading or taking compass bearings or in fact any aspect of the rally other than staying up late. I was secretly rather pleased about this, as it meant I could get on with it and lead us through the course quickly and efficiently. A friend had told me about getting someone to walk along the compass bearing you wished to follow while you stood still and then waiting while you went to catch them up, but I didn't think that kind of rigmarole was going to be necessary. Choose a route, stick to it, find the checkpoint, move on. We started out on our straight line.

As it turned out, though, there was no checkpoint down in that particular area of the woods—appropriately, near to a spot known as Comical Corner—and not only did we drift off the line I had set for us to follow, as there was nothing to find there was no intercept point, so that had we stayed on course we wouldn't know how far we'd gone in any case. Throw in the fact that it was night time, pitch-black, bumpy, tree-filled, a terrain of maximum obscurity even when visible, and I'm sure you can guess what happened. Within ten minutes— five, in all probability—we were hopelessly lost. Well, I was, anyway. My *teammates* were so uninterested in their location they didn't even notice, let alone get scared. They left all that to me.

CHAPTER 12

Loughton Hall | Lopping Hall

I never really told you about Loughton Hall, the home of Mary Wroth. It's down on the borders of Debden, the eastern flank of the town of Loughton, though the other side from the campsite. The building is set back from a road which acts as a release valve from the M11, squeezed in between a very ugly, tatty-looking, 1980s-vintage health centre and the newly rebuilt Epping Forest College of Further Education, the successor to the place where Penny taught art back in the 1960s. Debden was constructed after the war as a new estate to house people bombed out of the East End. Except for the row of oaks measuring out the road, the infrastructure of the whole area is beset by a kind of post-conflict plainness, where not falling down is the planner's only real aesthetic aspiration. To get to the Hall you squeeze down the narrow drive next to the health centre, which is mirrored by a tall hedge opposite so that you feel trapped between them, with nowhere to go if challenged. The lane opens out on to a forecourt full of cars and a large Victorian house whose main distinguishing feature is its impressive chimney stack, though the expert on such matters, architectural historian Nikolaus Pevsner, declares it 'an excellent building'. It was sold to London County Council at the end of the Second World War along with the land on which all the houses were built, and for a long while it was used as a community centre—teenagers dancing to records stacked on a Dansette, yoga classes, that kind of thing. At some point, the Hall became the responsibility of the FE college and in 2007 it was sold back into private hands—not without some controversy, as it involved the local council amending the covenants for the possible uses of the building. It is now an old people's home.

The first time I went to Loughton Hall I lurked right at the edge of the car park, took a bad photo and fled, almost as if I expected burly orderlies in scrubs to appear and shake their fists at me or confine me in a padded holding cell. I think by this point we have established my fear of trespass, of getting into trouble, of *being told off*. I wanted to go and have a look inside, though, so I found an email address for the manager and asked her if I could come and visit. It must have sounded like some kind of local history project and, in a way, I suppose it was. She was very businesslike about it all, but very helpful too. There was no reason she should be. But despite this invitation, or perhaps because of my own sense of imposture, I couldn't shake off the feeling that I shouldn't be there.

The day of my appointment was hot and I was running late. I scuttled to the meeting sweaty and dusty, much more nervous than the situation demanded. The car park seemed bigger than ever, the chimney stacks taller still. And the door! The door was an imposing, Victorian type of door, the kind that immediately has you searching for a tradesman's entrance and using spit to slick down your hair while checking your clothing for straw. A proper visitor turned up with his daughter as I stood waiting. I forget now whether he was the husband of someone who worked there or the son of a resident. Either way, it felt like he had more right than me, and also led to a moment of confusion when the door was answered, this mishmash group standing there firing off their mishmash reasons for being imposed upon by that portal. Then I was inside, in the quiet dark of the hallway, my figurative cap in my figurative hand, my figurative heart in my figurative mouth, while the manager finished up a phone call.

One of the reasons we know so little about Mary Wroth is because, her biographers believe, she left all her papers at Loughton Hall when she died. And, on 11 December 1836, almost two hundred years later, Loughton Hall burnt down. A servant woke at five in the morning to hear the bells that had recently been fitted in the anteroom, library and saloon all ringing violently. He rolled out of bed cursing his masters for summoning him at such a time only to find that the sound was created by the hot updraught of flame. Reputedly, the fire served as direct inspiration for an episode in Dickens's *Barnaby Rudge*. Here, the same sound heralds the burning down by rioters rampaging out from London of a great house he calls The Warren: 'The bell tolled on and seemed to follow him – louder and louder, hotter and hotter yet. The glare grew brighter, the roar of voices deeper; the crash of heavy bodies falling, shook the air; bright streams of sparks rose up into the sky; but louder than them all – rising faster far, to Heaven – a million times more fierce and furious – pouring forth dreadful secrets after its long silence – speaking the language of the dead – the Bell – the Bell!'

Fire engines were summoned from Romford and word was

sent to London for more, but at best they slowed the progress of the flames through fifty rooms while family, servants and locals rescued the bronzes, most of the paintings, the harp, the piano, jewels, clothes and all the plate. William Whitaker Maitland, the owner of the house, was either a bibliophile or a show-off: the library, newly refurbished, held over ten thousand books, plus family papers and, possibly, Mary Wroth's papers (though at the time she was of little interest to anyone). None of this was saved. Maitland's wife Anne wrote to her mother three days later and said, 'The Library is still burning or smoking…Every book, alas is gone. The Library which was just completed and perfect, the Study room, Dick's work baskets and boxes, every paper Lever valued and all my recollections of early childhood are alike gone.'

Flames are a recurring motif in Wroth's poetry. We can see them as the flames of love or the fire heating alchemical substances. The effect is much the same; that is, ambiguous. Wroth represents them allied to her belief in constancy but also as a kind of pain: 'although, itt pierce your tender hart / And burne, yett burning you will love the smart.' In fact, every theme in Wroth's writing is drenched (or burning?) with ambiguity. She can write, 'Cleere as th'ayre, warme as sunn beames, as day light, / Just as truthe, constant as fate, joy'd to requite.' But in her more saturnine moods, she thinks only of darkness: 'Night, welcome art thou to my mind destrest / Darke, heavy, sad, yett not more sad then I.' The ambiguity is not an accident, but her mode of expression, her view of the world. 'Leave the discourse of Venus,' she counsels elsewhere, 'and her sunn / To young beeginers, and theyr brains inspire / With storys of great love, and from that fire / Gett heat to write.'

•

The manager asked one of her staff to take me to the garden while she finished up what she was doing. They had a couple of tables with umbrellas out there, plus a bench along the wall. The first table was taken by a group of female residents who smiled at me and said hello as I walked past. A single old man sat at the second, his jaw working with a kind of absent intensity, nothing to chew on but his tongue. I went and sat on the bench, still sweating from the hurried walk, squinting under the full force of the sun. The back of Epping Forest Sixth Form College, looking like the company headquarters of a middling logistics firm with delusions of grandeur, sat uncomfortably close, the fence separating the two properties maybe twenty feet from the back door. It made the scale of everything feel off, as if college was advancing on hall under cover of darkness, night by night, inch by inch. One of the women at the first table came over to me. When she spoke it sounded like she had two sets of false teeth in. As a result, I wasn't sure what she said, but she held out a little box to me. It contained a rusty screw and a note explaining that she had excavated it on the far side of the garden during her own archaeological dig. Judging by her laughter and the laughter from her table, this was a joke, but at whose expense I wasn't sure, so I smiled and laughed one of those laughs which hovers and dies without ever landing.

Out came the manager—tall and authoritative, calm—and sat with me. I don't remember whether it was she who told me that the staff believed there was a ghost in the building, or whether one of them was outside dealing with some of the residents and shouted across to me. The nurse certainly joined in. She was very good-humoured about it, loud and funny, but completely in earnest. It was in one of the rooms upstairs, number eleven I think, or number fourteen. When I laughed the manager said she could see I didn't believe in that sort of thing, but it was clear from how she said it that she probably did, at least within the context of running an organisation whose staff believed in it. Actor-me is a very bad reporter—he was hot and uncomfortable and didn't make any notes—but I seem to remember that the ghost was a woman.

I t was the Maitlands of Loughton Hall who were largely, though accidentally, responsible for the establishment of Epping Forest as a park. In 1845, the General Inclosure Act was passed by Parliament. This meant that rather than having to pass an Act of Inclosure in the House for every separate case of Enclosure, local commissions took up the work of oversight and assessment of competing interests. Unfortunately, the Inclosure Commission thought its role was to promote Enclosure and, as Lord Eversley—who was to become intimately involved in the transformation of Epping Forest— notes, 'in its practical working [was] almost as detrimental to the interests of the labouring people' as the previous system. Between 1845 and 1869, almost six hundred and fifteen thousand further acres of English land were enclosed. That's over one thousand square miles. This wider trend was evident in Epping Forest. Charles II was the last king to have hunted on the land and the lack of royal interest in the forest had led to an almost complete collapse of the Court of Attachment, which was meant to protect vert and venison. This absence of authority had resulted, from the late eighteenth century, in a piecemeal Enclosure of small stretches of the forest.

The process was accelerated by the marriage of Catherine

Tylney-Long of Wanstead House, a now-vanished stately home near to where Old Mick and his fellow roads protesters would try, two hundred and fifty years later, to save an ancient sweet chestnut tree. In 1812, Tylney-Long was wed to the Duke of Wellington's brother, William Wellesley-Pole. As Wellesley-Pole's obituary put it, having been 'a spendthrift, a profligate, and a gambler in his youth, he became a debauchee in his manhood'. He burnt through his wife's fortune at such a rate that eventually Wanstead House had to be knocked down and sold for masonry. He also inherited the family post of Lord Warden of Epping Forest, appointed his own solicitor to head up the Verderer's Court and enclosed and then sold the four or five manors he had control over. By 1848 around a fifth of the forest had gone. That year a committee of the House of Commons recommended the intensification—or, at least, the ratification—of this process. Hainault Forest was to be enclosed and Epping Forest disafforested, with the Crown selling its rights to the local lords of the manors. The Act for Hainault was passed in 1851 and in a few short weeks the trees were pulled up and four thousand acres of land—an area two thirds the size of Epping Forest as it exists today—became fields. Rental income for the land jumped from £500 to £4,000. Noting this, some of the manors in Epping Forest quietly began to follow suit. Meanwhile, the Commissioner of Woods and Forests started selling off Crown rights on 3,513 acres at £5 an acre. By 1851, the size of the forest had dropped from nine thousand acres to six thousand.

In 1858, as part of the Crown's sell-off, William Whitaker Maitland bought 1,377 acres for £5,468. When he died in 1861 the estate passed to his son, the Reverend John Whitaker Maitland, who set about enclosing most of this new acreage. As the author Sara Maitland (no relation) puts it, there was already 'a considerable tradition of hostility' between landowners and commoners around the forest, but this latest Enclosure, so near to Loughton, caused particular resentment. Thomas Willingale—who sold lopped wood as part of his business and who lived up on Baldwins Hill—was having none of it and, at midnight on 11 November (the traditional start of a lopping season that ran until 23 April) headed out to Staples Hill with two relatives, broke down fences and gathered his wood as usual. Reverend Maitland first took the three men to court for injuring trees and, when this failed, for trespass. This time he was successful and, after refusing to pay damages, the trio were imprisoned for one week. The following November, 1866, Willingale and two of his sons—or perhaps his two sons and a nephew—repeated the action. On this occasion they were sentenced to two months' imprisonment with hard labour. At least one of the magistrates who passed judgment had received land from the Enclosure, although it's worth noting the ambiguity of Willingale's own position, as he was said to have unlawfully enclosed land round his own cottage. During their imprisonment, the story goes, one of the sons caught pneumonia and subsequently died. There was now very little room for anyone to back down, but at the same time there was little chance of Willingale, an illiterate labourer, winning any kind of victory. This is where the Commons Preservation Society stepped in.

Started by Lord Eversley in 1865, the Commons Preservation Society counted John Stuart Mill, William Morris and the subsequent founders of the National Trust among its number. This bunch of worthies had become increasingly concerned at the speed and extent of Enclosure over the preceding fifty years or so, particularly near to London. Within a few months of forming, the society was already involved in protests and court cases in Hampstead, Berkhamsted, Plumstead, Tooting, Wimbledon and Wandsworth, but the case of Epping Forest, because of the sheer size of the land involved, would be the most high-profile of all. The society contacted Willingale and told him they would put up a thousand pounds and fight a case on his behalf to prove his common right to lopped wood. The immediate effect of Willingale's agreeing to this proposal was an injunction to stop Maitland removing any trees or starting any building work on the land. It's hard not to see Willingale as a pawn in the game being played by these liberal toffs, but perhaps he didn't care so long as he was making life difficult for the local toff, Reverend Maitland. His case dragged on for a few years, refined and broadened in various ways, until he dropped dead in 1870. The society needed a new litigant, one willing to impeach the general Enclosure of the forest rather than just Maitland's. That would require a rather rich commoner. That would require the City of London Corporation.

I t's hard to describe exactly what the City of London Corporation is. It self-defines as a local authority, though it is also happy to admit that it's unique. Its territorial responsibility is for the City's Square Mile but its influence stretches much further. It has twenty-five electoral wards, only four of which are elected by its nine thousand residents, the rest by the corporations who are based there, the CEOs being offered more votes based on the relative size of their company. To become an elected representative you have to be a freeman of the City of London. This is most likely granted by one of the City livery companies, which in turn grew out of the medieval guilds.

Once that status is achieved you can stand for office, working your way up from common councilman to alderman to sheriff to Lord Mayor, each level responsible for electing the next. The Corporation's representative, the Remembrancer, sits behind the Speaker's Chair in the House of Commons in order to protect the City's rights and privileges. Remembrancer! A fantastical name for a man whose job it is to remind Democracy not to get ahead of itself when dealing with Money.

If the Corporation is a local authority it is, in effect, a local authority for the financial sector. It has no charter or constitution but states that part of its aim is '*officially*, not only to protect and promote the City's financial services, but also to promote financial freedom and liberalisation, *and* to fight those battles around the world'. Politicians have tried to reform or democratise it and failed. As Clement Attlee put it, 'over and over again we have seen that there is in this country another power than that which has its seat at Westminster'. It runs the Old Bailey courts, the Barbican Arts Centre and the Animal Reception Centre at Heathrow. Alongside the Church of England, in 2011 it was responsible for evicting the Occupy protesters from outside St Paul's Cathedral. The environmental activist George Monbiot has a memorable name for the organisation: Babylon. Father William Taylor, who has fought a long rearguard action against (and within) the Corporation, goes even further: 'It is a very intelligent demon, a dangerous thing...Maybe some things are best not talked about. I do think that it is spiritually very dangerous: the Corporation is a very dangerous place.'

In the 1860s the Corporation had acquired two hundred acres of land in the forest which it intended to make into a cemetery. When it was suggested to some of its officers that it could take up the Willingale case, 'this body, with a keen eye to their advantage', Eversley recalls, 'perceived that great popularity might be achieved by fighting for the interest of the public'. It's hard not to imagine that there was some measure of revenge against the lords of the manors at play here too. Until it was abolished in 1858, the City's Easter Hunt had been a source of considerable amusement to the local gentry—drunken city folk barely able to tell one end of a horse from another attempting to chase after a fat stag dressed in ribbons and released specifically for the occasion. How satisfying it would be, then, to exact a price from them. It's also true that there was a certain zeal for reform in Parliament in the late nineteenth century. In 1884, a Bill was introduced to merge the City and Greater London into one authority. That year's Lord Mayor's Show was particularly lavish and the biggest banner of all read, 'London would not be London without the Lord Mayor's Show'. Saving the forest served a similar purpose.

The case rested on the matter of who held common rights in the land—and in exactly what areas of land those rights could be claimed. The lords of the nineteen manors of Epping Forest argued that any common rights were granted by them or the Crown to the residents of that manor only. Hence, if they could reach an agreement with the majority of residents within their manor, they could enclose. Eventually, though, the Corporation and the Society proved that 'instead of being a congerie of separate manors, the Forest was one great waste, over which the Commoners of every one of the nineteen manors had the right of turning out their cattle, without obligation of confining them to their particular districts'. That being the case, it would be far too complicated for any man to show that he had the consents necessary for Enclosure. Hence the land could not be enclosed.

In 1874 the Master of the Rolls ruled for the Corporation, granting an injunction prohibiting any further Enclosures and requiring the lords of the manors to remove all fences erected in the last twenty years. The following year a Royal Commission was established to decide on the future of the forest. It found that Epping Forest consisted of six thousand acres and that over three thousand of them had been unlawfully enclosed, also that Loughton's lopping rights were valid at law. However, the final report in 1877 backed off from the logic of its own conclusions and tried to suggest a 'compromise' whereby the lords of the manors could keep their fenced-off land. George Burney, a member of the Commons Preservation Society as well as a landowner and commoner in Epping Forest, decided that the best way to oppose this proposal was through direct action. 'With the aid of a large body of men,' Eversley writes, 'he forcibly removed the fences from many of the inclosures'. He doesn't ever say it, but by this stage the situation was out of the hands of the society. While the politicians argued, the Corporation had been buying forest land: up to three thousand acres at around twenty pounds an acre. While this was nowhere near what the land was worth for development, it was a high price for waste land—and four times what many of them had paid when buying from the Crown—so owners were keen to cash in.

The battle was over, another crisis resolved. The Epping Forest Act was passed in 1878, stating that the forest should remain open and unenclosed for all time for the recreation and enjoyment of the people. The Corporation would control and manage this national resource, named in the Act as its Conservators. Crown rights and Forest Law were abolished and all land enclosed in the last twenty years was to be restored unless already built on. But here's the thing: the Act also ceased all lopping rights, without compensation, and when Eversley tried to amend this clause, it was the Corporation, the supposed saviour of common right, 'very unfairly, as I thought', who opposed it, threatening to go to law and argue much the same case as that previously put forward by the lords of the manors. Eventually they lost the case: lopping was to be abolished, but the Corporation would have to pay compensation. Seven thousand pounds was set aside. One thousand went to the families who made some of their living from lopping and the rest was used to build Lopping Hall in Loughton. This community centre still stands today, rather scruffy on the outside but large and fully functioning within. I went to watch some wrestling there. The son of two northern punks who had moved to east London to be nearer to Dial House was taking part in his last ever bout as Dickensian Docklands villain Shaun Sikes. The main room was just about big enough for the ring and, despite some enthusiastic booing, I almost cried when he limped out, defeated.

Under the terms of the Epping Forest Act, Reverend Maitland received £30,000 for the land his father had bought for less than £6,000 twenty years earlier and, suddenly flush, he immediately commissioned the Gothic Revivalist architect William Eden Nesfield to rebuild the burnt-out skeleton of Loughton Hall. It didn't work out too badly for Maitland, all things considered. The effects of stopping lopping, though, are still being felt today. The Victorians saw the forest as something existing independent of human interference. As far as they were concerned, lopping just created ugly, stunted trees and ugly, stunted people. But, by keeping the tree canopy thinner and letting more light in, lopping supported a whole ecosystem on the ground beneath the trees. Now huge areas of woodland have nothing growing on the ground at all, just a carpet of dead leaves, and the Corporation's staff are trying to relearn and, in a few places, implement the skills banned one hundred and thirty years before. Epping Forest is neither a traditionally 'farmed' forest any more, nor a wild, high forest, and yet it's both.

There is one utterly beautiful sonnet in Wroth's work. In it she sets out how trees mourn for the loss of summer: 'Thus of dead leaves her farewell carpett's made: / Theyr fall, theyr branches, all theyr mournings prove; / With leavless, naked bodies, whose huese vade / From hopefull greene, to wither in theyr love.' In a final two-line stanza, she delivers her killer punch: 'If trees, and leaves for absence, mourners bee / Noe mervaile that I grieve, who like want see.' It may not be an original idea, but that end, that unpunctuated defini-tion of humanity as those 'who like want see' is so simple and opens up such pathos, that I found myself going back to it, reading it for exactly that way in which the formal tweak—the lack of commas—creates or amplifies its meaning.

Wroth is obsessed with the concrete reality of words on a page, with *form*. In some ways this goes without saying. The sonnet is a container for words and ideas and also something in which they are contained. All poets of the early modern era were formalists, in that they all worked with restrictive forms. Wroth's poetry, moreover, with its strong emphasis on tradi-tional, and extremely taxing, Petrarchan rhyme schemes, was seen as a little old-fashioned and restrictive even in her own day. But let's focus on the punctuation in this sonnet. She uses eleven commas within the fourteen lines of the poem, plus six semicolons, two colons and two full stops. She is neither afraid of punctuation nor ignorant of its function, yet she opts at the very end for 'like want see', running the words up against each other so that they become almost a composite and humans, in turn, become trees who likewantsee.

A related punctuational quirk is evident in Wroth's prose work, where she is consistently suspicious of the full stop. As the editor of the most recent version of *Urania* points out, 'Wroth's unit of thought is the clause rather than the sentence. Like her narratives and perhaps like thought itself, her rush of additive clauses defies closure.' And she isn't the only one of the subjects of this book to display this wanton disregard for the rules of grammar and punctuation, this resistance to (En)closure. Ken Campbell wrote all three of his one-man plays using a hyphen at the end of each line, in honour of the collector of weird phenomena, Charles Fort: 'It's Fort's opinion that everything in the Universe is linked with everything else — so a Full Stop is a Lie — Or a Hyphen coming straight at you.' John Clare was roundly ashamed of his lack of punctuation, seeing in it his humble origins, but now readers accustomed to modern free verse find only power and a lack of affectation in his uncorrected originals, ambiguities multiplying in his blocks of text. I read somewhere that Penny Rimbaud adopted a similar technique to Campbell's when writing the original version of *Christ's Reality Asylum*, using only forward slashes, but when I looked for the quote I couldn't find it. I emailed Penny to check and he replied:

there is probably a multitude of reasons for the dashes/ if I'm on record for giving an explanation/then/for certain/that explanation would have been but one of that multitude/here's another/it's a lot easier not to have to consider conventional abbreviation or grammatical laws/which/as I'm sure you know/pose all sorts of difficult questions/not least the ever worrisome comma/ which on the one hand might have been laziness on my behalf/or on the other/a refusal to accept academic authority/

When I told a friend, sitting in a pub, that I intended to go and spend a night out in Epping Forest, he said I was mad. He was very certain about it. He didn't deny that it was a good idea in theory, just told me that it was a bad idea in practice. Once I explained why I needed to do it—easy in the way that anything seems easy after a couple of pints—he tried to make me focus on the practicalities, already taking it more seriously than I was. I would need to sleep up a tree and I would need to wear night vision goggles or I would certainly be murdered. Despite arguing that wearing night vision goggles would only enable me to *see* myself being murdered, I staggered home later that night and looked for them on eBay. The ones from the former Soviet Union are massively expensive and need a huge battery pack, and I'm presuming the version you can get for a tenner simply don't work. Moreover, to be able to see everything that was going on around me would rather defeat the point of the exercise. I decided he was right about the tree, though. It fitted with my repeated, though inept, attempts at boling so neatly, so perfectly that it felt less like real life than the plot twist from a cheap novelisation of my existence.

The problem I had was finding the right tree. We've established by now that I was neither very good at getting up trees nor completely comfortable with getting down. To this already disappointing complex had to be added the matter of *staying up* a tree, indeed *staying up it all night*. I needed somewhere not too high, though high enough to offer protection from roaming lunatics and crazy foxes. I needed it to be easy to get up and then easy to sit in but at the same time to be discreet and hidden, somehow. Basically I needed a magic tree from Enid Blyton, something as anodyne as possible.

I began by looking for thematic unity, determined that somehow, in some way I didn't really believe in, it was all meant to be. I went back to the Skull Tree to see if I could scramble up and get comfortable in there, but the branches on that one all went straight up very close together. I strolled from there to the Lovely Illuminati Tree, thinking of Ken, but while I could climb it, I still felt I'd lurch out of front or back at any moment. In addition, both it and the Doom Tree sat on a busy path and the sheer weight of carvings they were carrying suggested that they might be some kind of nocturnal meeting point for local adolescents armed with drink, drugs and the requisite knives. The Cloven Tree was another I couldn't sit in, nor get out of with any remaining self-respect. I kept looking.

The radical indeterminacy of Wroth's work runs deep. Both books of *Urania* 'finish' with incomplete sentences. The first part ends, 'Pamphilia is the queen of all content, Amphilanthus joying worthily in her; and' while the second goes for 'Amphilanthus was extremely' as its closer. To put this in the terms we employed earlier, it's as if Wroth doesn't see the aim of alchemy to be the creation of a philosopher's stone, but that the continuing act of alchemy *is itself* the philosopher's stone. This emphasis on process over outcome reaches its peak with her crown of sonnets. The crown or corona would have been a form she was already familiar with. Its roots lie with the Petrarchans, but her uncle, Sir Philip Sidney, included one in *Astrophel and Stella*, John Donne wrote his religious sonnet sequence *La Corona* around 1607 and even her father, Sir Robert Sidney, had a go at one, though he failed to complete it.

A corona is supposed to crown the sonnets that have come before by adding an additional, taxing constraint. The last line of each sonnet becomes the first line of the next and the poems curve round so that the last line of the last sonnet is the first line of the first. While Donne's and Sir Philip's sequences were shorter, and her father didn't even get that far, Wroth goes the whole hog with fourteen. And with the first (and last) line she turns her poems into architecture, a structure for your mind (and hers) to wander in: 'In this strang labourinth how shall I turne?'

In this strange labyrinth how shall I turn? The structure means, quite clearly, that it doesn't matter which way she turns. She is trapped and she (and the reader) will find herself back here, asking this same question again and again and again. She will start over, full of hope, and try to follow through the maze 'the thread of love', which she equates with chasteness, constancy and faithfulness, but at some point—exactly where it's hard to tell—it will all start to go wrong. She encounters a degree of indifference from the subject of her love, perhaps because he doesn't love himself and is 'longe frozen in a sea of ise'. Lust makes an appearance, a 'monster', the changeling baby of love, and we find ourselves very quickly contemplating the 'Fruit of a sowre, and unwholesome ground'. Before we know quite how we got there, the talk is of hemlock and ashes and 'those cool, and wann desires'. Wroth rallies a little, the fog clears, she praises wisdom, 'holly friendship' (hopefully not the bush) and truth before committing herself to the Great King of Love. But then right at the end, when she seems to have rescued herself, it all goes wrong. 'Curst jealousie doth all her forces bend / To my undoing; thus my harmes I see. / Soe though in Love I fervently do burne, / In this strange labourinth how shall I turne?' And we begin again. There is no escape. There will never be any escape. Romantic love is a trap. If it is a story we tell ourselves, or about ourselves, it is circular and never-ending. It has no real beginning or end, so it can't have a middle either. We are lost in it and imprisoned in it.

But in what sense is the labyrinth strange? Towards the end of *The Tempest*, Prospero gathers together his protagonists and shows Alonso, King of Naples, that his son is alive and well (and incidentally in love with Prospero's daughter). Straight after, Alonso's boatswain enters to tell him that their vessel, shipwrecked and destroyed at the start of the play, now miraculously rests at anchor, ready to set sail. Alonso exclaims, 'This is as strange a maze as e'er men trod, / And there is in this business more than nature / Was ever conduct of.' It seems almost unimaginable that Wroth would not have seen *The Tempest*, which was first performed before James I and his court in November 1611 and revived the following winter as part of the celebrations of the forthcoming marriage of Elizabeth Stuart and Frederick V, the Elector Palatine. Perhaps this was the seed from which her own line grew. In which case, there is an implication of magic, of bewitchment, embedded in it.

The *Oxford English Dictionary* lists Wroth's original spelling, strang, alongside four meanings only. The first is as part of the phrase *strange woman*, meaning a harlot, a translation of two different Hebrew words in the Bible, strictly speaking, *not one's own wife*. The next two are *added or introduced from outside* and then *alien, far-removed*. The last is *unknown, unfamiliar; not known, met with or experienced before*. The labyrinth is an alien structure, even if in some compacted way, that labyrinth is Wroth herself—*not one's own wife*. If it is her own body, she is radically transformed. She, or it, is *new*.

In *Caliban and the Witch*, the radical feminist academic and activist Silvia Federici argues that the witch hunts of the sixteenth and seventeenth centuries were a necessary part of the birth of capitalism, as more traditionally represented in the primitive accumulation of Enclosure. 'The witch-hunt was one of the most important events in the development of capitalist society...the unleashing of a campaign of terror against women...weakened the resistance of the European peasantry to the assault launched against it by the gentry and the state, at a time when the peasant community was already disintegrating.' Breaking this resistance was important, but perhaps even more central to the establishment of capitalism was positioning women predominantly and primarily as (re)producers of labour, that is, the creators of a workforce. In order to do this it was necessary to break down any status they had within their society as keepers of wisdom or knowledge, much of which was in effect outlawed as witchcraft. As we saw earlier, Descartes's friend and associate, Marin Mersenne, cleared the way for the former's ideas by connecting the Renaissance thinkers Wroth would have respected to witchcraft and dangerous magic. Indeed, Federici argues that only by breaking the idea of the body 'as a receptacle of magical powers' and instead turning it into a Cartesian meat-machine for work were the conditions for capitalism fulfilled. It was Francis Bacon who said that 'Magic kills industry', just as common land did.

What if Wroth's corona is itself encoded, so that enclosed within the surface meaning there is another? And what if this meaning is alchemical? The heating and cooling of elements, their bonding and breaking within a hermetically sealed vessel (the labyrinth, perhaps strange precisely because it has no exit) is certainly consistent with this reading, but where does it take us? In the Borges story we touched on earlier, 'The Garden of Forking Paths', the character Stephen Albert discusses 'the ways in which a book can be infinite. I could think of nothing other than a cyclic volume, a circular one. A book whose last page was identical with the first, a book which had the possibility of continuing indefinitely.' Twice *Urania* ended without ending. Wroth's crown of sonnets takes the process a step further. By making the process of alchemical transformation infinite and cyclical, Wroth demonstrates as clearly as she can that it is indeed the process—this ongoing, relentless process—which is its own purpose.

Magic never ends. It's not only hard work, it's hard work which goes on forever. The result of it, apparently, is to wind up back where you began. Magic never ends and it has no product and *this* is why it's inimical to capitalism. There's a link between the idea of the Cartesian self, the developing notion of romantic love, the notion of progress, the attack on witchcraft, increasingly structured and heavily reinforced gender roles and the birth of capitalism represented by Enclosure. Magic doesn't need to fight these changes or stand in opposition to them—it short-circuits them or renders them meaningless, not by dint of what it produces but by dint of what it is. It's not only *not* a process to manufacture gold (that's capitalism), it's not a process to manufacture anything.

Wroth offers us radical ambiguity as a charm to pull up the fences of Enclosure, an acid to break down the raw matter of 'reality' before re-forming it and then breaking it down, again and again and again. As noted earlier, it's believed that *The Tempest* was performed during the betrothal celebrations of Elizabeth Stuart and Frederick V, but also that the masque towards the end of the play may have been added particularly for this occasion. When it's done, Prospero tries to reassure his future son-in-law: 'These our actors, / As I foretold you, were all spirits, and / Are melted into air, into thin air: / And, like the baseless fabric of this vision, / The cloud-capp'd towers, the gorgeous palaces, / The solemn temples, the great globe itself, / Yea, all which it inherit, shall dissolve.'

There's a problem, though, if we're going to treat all this as a reason to view the forest as some kind of zone of liberation. Epping Forest is ruled by a demon, an ancient, unelected organisation committed to exactly the kind of *liberalisation* which enclosed the land two hundred years ago and crashed the world economy in the last ten, the kind of Thatcherite ideological powerhouse which crushed the Left in the '80s. The 1880s in this case, or the 1780s, or the 1380s, when the Lord Mayor, William Walworth, killed Wat Tyler (an event commemorated by a stained-glass window back at the Corporation's headquarters, Mansion House), or the 1080s or 880s or some time back before anyone had even considered the idea that *liberalisation* was a word for corporations being allowed to do whatever they liked; a word for allowing and encouraging all restrictions of access to space—physical, intellectual, psychic, emotional—brought about for private gain. None of this should be surprising. The forest was king's land, *forestare* means to exclude. The forest itself was a kind of Enclosure, though one full of gaps and unintended benefits. And how fitting, really, that control of that land has passed to the kings of Finance, of the free market. The only possible response comes from Calvino's baron, not because it's literally true, but because of what it means when you say it (like a spell): 'it's the ground that's yours and if I put a foot on it I would be trespassing. But up here, I can go wherever I like.'

The manager showed me right round Loughton Hall. I don't remember that much about it. It was dark in the hallway in that classic Victorian style, the wood stained almost black. She took me upstairs and showed me a corridor and some old pictures. We went out the back to where the stables had been converted. She led me through the two main rooms, both of them lined, along every wall, by old people in those comfortable, wipe-clean chairs you only get in hospitals and care homes. They all smiled at me as we passed through, some talking to her, some talking to me. My hearing seemed to go off-centre so I struggled to make out what anyone said. It was an orderly, light, happy place, the kind of home you'd willingly leave your mother in. The staff kept telling me about the ghost, a woman or a girl, I don't know. They were larking around, certainly, but they weren't joking, and in a way it wasn't hard to see why they believed in ghosts. The place was full of them.

When we were sitting on the bench outside in the sun the manager explained to me that the Hall was originally conceived as a general old people's home with a dementia specialism. But all the patients here had dementia at one level or another, she told me. Demand had proved to be even higher than they'd expected. 'Alzheimer's is…a sort of mental confinement,' writes the journalist and author David Shenk in his book on the illness, *The Forgetting*. 'The sufferer is incarcerated within the collapsing neural structures that he has taken a lifetime to build.' But dementia is a little like getting lost too, a strange labyrinth where what you've lost is yourself and you wander the passages of that self looking for it without ever realising you're in it. And towards the end it's a lot like being dead, even while you're still breathing. Ghosts, forgetting, incarceration and another—very final— way of losing yourself. A haunted house packed with things to frighten the irresolute. In this strange labyrinth how shall I turn?

Jacob Epstein's last two surviving children, from the two sides of his bipartite family, died within four months of each other in the winter of 2010–11. Peggy Jean Hornstein Lewis was the first to go, aged ninety-one. Her last days were spent in the Dosher Nursing Home in North Carolina, part of the hospital that her husband Norman—the beautiful Jewish-American communist doctor—had worked in from the 1950s until his own death. In an echo of the methods of Dr Allen's forest asylum, 'the nursing center encourages residents to care for a courtyard garden, birds, plants and flowers so they may become active members in their care and become as independent as possible'. Peggy Jean seems to have devoted her life to a kind of cheerful normality and 'she…leaves behind many others that loved her joyful, giving and humorous spirit, particularly the Dosher Nursing Home family'.

Her half-sister, Kitty Godley, the second of Kathleen Garman's three children, died the following January. Godley had been married to Lucian Freud after the war and is the subject of his remarkable series of early paintings including *Girl with a Kitten*, an intense and sensuous dream-logic portrait held together by Kitty's huge brown eyes, looking off to the right, and the green eyes of the cat, looking straight out at the viewer. Kitty was both extremely sensitive and a survivor. You would need to be to deal with a mother like Kathleen Garman. 'I think she wanted her daughters to excel,' said Kitty, 'but she didn't want us to succeed, because she had to be the queen.' Appropriately somehow, her second husband, who died earlier the same year, was an oboist who was too scared to play in public and who instead became an economist. Like Peggy Jean, she was remembered, above all, for her warmth and kindness. 'The dementia of her final years,' the author and Garman expert Cressida Connolly writes, 'did not diminish her gentle sweetness.'

The tree I found was further south than I'd expected, not in any area of the woods that I associate with the biggest, oldest specimens, quite near to the clubhouse of the cross-country club my son runs with. It wasn't that far from Chingford, maybe a mile north-east of Pole Hill, where Vyvyan Richards had his retreat. I was pushing through a whole irritation of holly bushes one morning when I blundered out onto the edge of a clearing where forest workers had thinned back the trees in line with their new policies. Right in front of me, as if put there for my purposes alone, was another beech, three massive branches shooting up from its historic pollard, a particularly harmonious, symmetrical and simple specimen. It had footholds at just the right height and the short stump of a removed branch in the perfect place to pull up on. There was a hollow about ten feet from the floor that I could sit in with my back against one branch, the other two on my left and right and a smaller, younger protuberance sticking out where I could rest my feet. It was perfect. Behind me were thick holly bushes, in front a cleared area so that I could even see some sky above it, plus any advance of murderous hordes. Beyond the clearing ran a path which would return me to the main road in less than ten minutes if I lost my nerve and had to flee. The only marking on the tree was a couple of metres above the hollow on the left branch, where someone had carved 1966 and underneath 1968. Apparently they hadn't been back since. I sat there, anchored in wood, armoured by the branches on each side, and felt an obscure calm creep over me.

CHAPTER 13

Orme Square | Covert Woods

Narrator-I and actor-me—manipulator and goof—came together again at the National Archives in Richmond, south-west London, crouched over court papers for two cases involving Old Mick. Both the folders were fat, replete with statements, a couple of photos and the scribbles of the lawyers who had worked through them. Reading them elicited an odd, uncomfortable feeling. On the one hand, this was direct history, everything in front of me, from requests for legal aid to a handwritten letter sent by Mick to the judge in a case while remanded in H. M. Boys' Prison Wormwood Scrubs when he was seventeen. In the seven years between 1957—when he was still only fifteen—and early 1964, Mick had compiled a long criminal record involving everything from stealing sweets to attacking a copper with a screwdriver, escaping from Ashford Remand Centre, receiving stolen goods and driving away a variety of motor vehicles 'without consent'. But right now I felt that it was me engaged in a kind of trespassing, not Mick. There was an uncomfortable feeling of being very close to people and events that had nothing to do with me, that

I had no right to comment on. Looking through these files, another facet of the man was highlighted, but I'd only found it by breaking and entering.

A t eleven thirty on the morning of 24 March 1964, Mrs May Harvey, the wife of print tycoon Wilfred Harvey, heard the doorbell go at 1 Orme Square, W2, a huge townhouse in a very exclusive area just north of Kensington Gardens. Mrs May, described later by the police as 'an elderly lady', headed down the stairs to the front door but there was no one there. When she turned to go and check the back door she found 'two men standing on each side of me—a short man and a tall man', a double act with no jokes. The men told her to get upstairs quick. They took her into her son's bedroom but she made a run for it and the three of them ended up in the bathroom. Here, they sat her on a stool. Seeing the jewels on her fingers, the short man told her to 'Take off that ring or I'll cut your hand off.' He ripped the necklace from her neck and, while the tall man held her on the stool, wrenched the ring from her finger. She got up and made a run for it again. Out on the landing they pushed her into a cupboard. It wasn't that deep and she was scared of being locked in that narrow space. The tall man 'took his big padded glove and put it over my face, pressing it hard onto my face…I struggled and hit out at him.' That was when she said it happened. The tall man 'punched me on the nose and on each side of my head— he must also have punched my chin'.

It sounds as if they were all a little taken aback after this. Mrs Harvey got out onto the main landing and asked for fresh air. She sat on the stairs and they brought her a drink of water and something to wipe her bleeding nose with. While the tall man guarded her, the small man picked up a stool and used it to bash in the door and then the back panel of the cupboard in Mrs Harvey's bedroom, behind which was the family safe. The tall man went to get the keys from Mrs Harvey's handbag.

Ivy Tyler, Mrs Harvey's domestic, arrived at around noon. Walking into the kitchen, she found a saucepan on the ring, burnt dry. The handle was hot and she let go of it fast, so that it clattered back on the stove. She was standing, wondering what had happened, when she saw the tall man coming down the stairs. He grabbed her by the shoulders and pushed her ahead of him: 'Quick, get upstairs.' Mrs Tyler was confused. 'What's happened? What's gone wrong?' 'Keep your head down,' the man replied, 'I don't want you to identify me.' She only realised what was happening when she saw the mess on the landing and Mrs Harvey with blood on her hands and face.

The tall man pushed them both back into the bedroom, ripped out the telephone in there and told them not to move. 'He said, "Stay there, the pair of you, any tricks and you will both get it." He was using filthy language all the time.' While this was all going on, the small man had used the key to open the safe and scooped its contents out into a pillowcase, a haul of jewels reputedly worth almost one hundred and fifty thousand pounds. The two women were left alone in the room for a moment, neither daring to speak. The tall man called through to them: 'If you identify me, if it takes me ten years I'll come back and get you.' And then there was silence.

The women didn't leave the room until they were sure the house was empty. They went downstairs and Mrs Harvey rang her husband and the police. Her doctor was summoned and arrived at 1.30 p.m.: 'She was very shocked and harassed; she had several bruises and cuts; her nose was bleeding – there was bruising and abrasions on her forehead; her lower lip was bruised. Her left ring finger…was swollen, red and cut. It looked as if it had been subject to a violent pull, and as if something small had been pulled off it. Her left forearm and wrist had scratching and bruising on them; her right wrist was also bruised. She was suffering from shock which was the worst factor in my opinion.'

The short man was caught first. William O'Gorman, twenty-three years old, was arrested at the start of May and around half of the stash of jewels was recovered. The officer leading the case travelled to Dublin and recovered more of the stolen items. In July O'Gorman was sentenced to seven years in prison. Just over a week later, Mick Roberts was picked up, staying at a house in Chelsea. He was taken to Notting Hill Police Station and put in a line-up. Ivy Tyler didn't pick him out, but things were different with Mrs Harvey. 'Almost before I had finished speaking,' stated the sergeant who ran the identity parade, 'she said, "Yes, I recognised the man straight away when I came in." She was looking directly at the defendant and indicating him with her arm. I asked her to touch him and she looked rather frightened and said, "Must I?" She then went straight up to the prisoner and touched his arm and said, "You recognise me, don't you?" Defendant who looked very dismayed, said, "No".'

The next day Detective Inspector Johnson interviewed Mick again and asked him if he wanted to make a statement. 'No,' he replied, according to the police, 'you've got me bang to rights – I've got no chance so what's the use? O'Gorman did a bunk with all the tom. I never had a bean out of it – I know they think I grassed – it wasn't me – it was Maple; he had worked there at the place and set the job up ... I suppose I'll get ten years because they have put the violence on me; but you can take it from me the woman wasn't hurt, although I stand no chance I must have a go for it.' He was right. He was convicted and got ten years. When his appeal against the sentence was turned down, *The Times* reported the judge's condemnation of this 'barbaric crime'.

I fell in love with my tree, precipitously, disastrously. I would find myself looking at its location on Google Maps when I should've been working. When my son went to his Wednesday-night runs through the woods I would sneak off and climb it and sit there as it grew dark, crows—sardonic— letting me know what they thought of me, the light leaving me so I had to jump down into blackness. I twisted my ankle the first time I did it, scratched my palms so they bled, but I didn't mind. I felt slightly ridiculous perched up there, but oddly peaceful, too, a big, ungainly puppet of a bird crouching in its nest. I began to entertain the thought that I might enjoy my night out here, that rather than being terrified and/ or attacked I might find myself *at one* with something, transported by beauty, in receipt of revelation. Though I couldn't bring myself to articulate what that might be. And with this calm, this belief in forthcoming epiphany, I guess I'd achieved what I hoped to achieve in some small way—I was bewitched.

Of course, part of this was a way to avoid the growing contradictions I found embroiling my project, the way that Either and Or seemed to be moving further away from each other. Narrator-I was trying somehow to hold the thing together, to *justify*. Actor-me was up a tree. I was researching Mrs Harvey's husband, Wilfred Harvey. It was reported in

327

Parliament a year after the break-in that he'd fiddled his taxes and it turned out that the firm he ran, the British Printing Corporation, 'suffered massive financial irregularities' under his leadership. I wanted there to have been a fit-up, a conspiracy of some sort. I wanted to say that the narrative presented in those files was a police narrative and hence put there in defence of the status quo (which, of course, it was). I kept talking down the magnitude of Mick's offence because of where it left me—and I was only talking to myself. I was guilty of doing it again and again. The truth is, there was something so trite and nasty about it all, about punching an elderly lady in the face, that it knocked some of the stuffing out of me. It didn't fit with the narrative I wanted to believe about Mick Roberts the glamorous 1960s cat burglar, or at least only with the very worst, the darkest and most hidden aspects of such a narrative.

The more Mick's account of his life didn't match up with what facts I could establish, the worse I felt. Mick's friend Russell, who I'd gone and visited, remembered Mick telling him that he had raised his hands and placed them on May Harvey's shoulders when she was *resisting* and that this alone was enough to inflate the charge from robbery to robbery with violence. Nikki and her mother thought he had taken the rap for something which wasn't his fault, something relating back to the Richardsons in south London. And that was just the start of it. Mick had told Russell and other activists that his front teeth had been knocked out while being force-fed to break a hunger strike. Looking back through my notes, I saw he'd told Nikki that he'd lost them in a fight at a time when he wasn't even in prison.

I managed to get hold of an email address for Nell Dunn, who wrote the novel *Poor Cow*, on which Ken Loach's film was based. I asked her if Dave was indeed based on Mick Roberts. 'The character of Dave in *Poor Cow*,' she replied, 'was more or less fantasy from talking to people and in the book he is rather a non-event, the important person being Joy. I may have met Old Mick in my research but I'm afraid I don't remember. (Old age and bad memory don't help.)' I also emailed a man who had written a number of books about the Richardsons and he said he'd never heard of Mick.

I pulled up old copies of *The Times* from when John McVicar did his bunk from a prison coach on Dartmoor, but Mick wasn't among the men listed as escapees. Nor, when I looked through the archives of the same paper from 1958 to 1975, was there any mention of a Michael Roberts making an escape of any sort from Dartmoor Prison. Indeed, of the twenty-seven reports I found where the escape method was listed, only one prisoner appeared to have got out over the wall unaided and it wasn't him. As *The Times* stated in 1964, 'the method of escape which has stood up best to test is from outside working party'—the groups of convicts taken onto the moor as labourers. Even the famous excarcerator, Wally Probyn, described in his time as a 'boy Houdini' and already the veteran of seventeen prison escapes, ran off from a working party a little later that year. Maybe he'd read that report in *The Times*. It was presumably true Mick had been in Broadmoor, for the simple reason that I couldn't understand why anyone would make that up. Why he was in Broadmoor was another matter. A cunning ploy to get himself out of prison quicker didn't seem like the most obvious explanation.

I felt uncomfortable, anxious. I was overcome with tiredness, waking in the mornings slowly and with difficulty, filled with a heaviness I was sure must have a physical cause.

My writing life had been spent trying to cheat my mind into getting on with the job instead of veering off into displacement activities, and one way I did this was by regularly changing where I worked. Recently I'd started renting a friend's garage in which he'd fitted a small wood-burning stove to keep me warm. In my mind it was supposed to be a simulacrum of a forest hut, a kind of capsule to trick me into thinking I was out among the trees. The stove smelt and I began to worry about carbon monoxide. I felt like I was being poisoned. Perhaps I was poisoning myself. I'd been trying to establish the truth of Mick's account and, in failing, I ran the risk of undermining other people's belief in him, their need for their own stories of him. I didn't think it was my place to intervene between a dead man and his friends and family. But then again, I couldn't let him go.

In retrospect, you could see my project like this. Finding myself in a dark forest of the mind—the dark forest Dante suggests awaits us all in the middle of our lives, though I suspect that each has its own terrain—I had externalised that metaphor onto an actual landscape, six thousand acres and more of real trees. Having done so, I began looking for stories about the residents of this landscape which spoke to my dilemmas as I saw them, less as examples of how I had lived my life so far than as pointers to how I could live it moving forward. And, as I had come to see my weaknesses in terms of fear, in terms of an unhealthy respect for boundaries, a craven deference to authority, an awful need to be liked, then Mick Roberts had seemed to me some wild god of boundlessness, if not a person I'd want to copy then at least a distant star, something to reset the sextant of my self by. But now all I could hear was Nikki telling me what she'd told her mum: 'You know what Mick's like. You can never believe what he says.' I was stuck, caught somewhere between truth and a good story.

I went to visit Susanna, another roads protester. She lives over in north London in a small council house down the back of Muswell Hill, much nearer to the North Circular than the big villas round Alexandra Palace. She was minding a neighbour's dog the first time I called, and he barked when I knocked and came bounding out to greet me and I had to go into my whole routine about how I was nervous with dogs and would she mind terribly holding him back? Once I was inside he stopped barking and kept bringing me some kind of silver moonboot as a peace offering, an apologetic set to his features, so I felt ashamed of my behaviour in the driveway, the squeakiness of my response. Susanna told me about her grown-up children and her grandchildren, including the grandson who is on Arsenal's books, all the usual stuff you'd expect from an intelligent, proud seventy-year-old woman. Like Nikki—and indeed, Penny and Gee—she didn't seem like she was seventy, which suggests my idea of how seventy-year-olds should look and act is lagging behind reality.

Susanna is a veteran of the roads protest movement. She was there at Claremont Road, at Kingston, at Crystal Palace, you name it. She took her kids out of school for a year to protest at Newbury, where they all lived in a bender. But she's more than that too, some kind of ultra-networker of British counterculture. Originally from New Zealand, she worked on the radical fringes of British theatre in the 1960s, so that when I mentioned Ken Campbell, she could immediately whip out the programme from his memorial service to show me. (For the record, I don't think she was massively fond of Ken, mainly because his attitude towards women was, ahem, a little bit *backward*.) She describes herself as 'a devout psychonaut', but when people tell her it's the acid speaking, she always replies, 'I was like this long before the acid, darling.' One time when I was there she said a friend of hers was popping round. He happened to be the ex-partner of a school friend of mine, an activist-squatter who was himself involved with the No M11 campaign. She knows Daisy Eris, Ken Campbell's daughter, and she had met and hung out with Robert Anton Wilson. She's one of those people who likes slotting other people together, making links, finding connections. The only other person I've known like this had also moved to England as a child. I wonder if that sudden developmental dislocation has something to do with it.

She is also an impossible interviewee, maddening, someone you chase along, barely able to get a word in, trying to herd the cats of her thoughts in the right direction while they scuttle off down ever more recondite paths of their own choosing. I'd gone there to find out about Mick, but she never told me much about him. She recounted the anecdote about his front teeth and force-feeding and a good one about how, if Mick was going to punch someone, he always pulled on one glove, so that whoever had crossed him knew to make himself scarce if he saw him doing it. What she told me about most, though, was King Arthur.

King Arthur is a former biker who, in 1984, decided or was informed that he was the reincarnation of Arthur Pendragon, King of England (or Albion? I'm really not sure about the exact nomenclature of ancient, possibly non-existent kingdoms). His story is brilliantly told in the book he collaborated on with the author and journalist C. J. Stone (another friend of

Susanna's who has also written about Wally Hope and Penny Rimbaud), so we won't go into it in too much detail here. It's probably enough to tell you that Arthur carries Excalibur at his belt, the one used in the film of the same name, and that he bought it for a couple of hundred quid from the man who made it after the guy had told him that he'd only sell it to the actual King Arthur, and Arthur had got out his driving licence and passport to prove it. Also, that he now leads his own order of druids, the Loyal Arthurian Warband (LAW for short), of which Susanna is a very high-ranking member, both Faery Queen and Battle Chieftain. He is also the Pendragon of the Glastonbury Order of Druids and the Swordbearer of the Secular Order of Druids. The druids take him very seriously. In addition, he and his knights 'fought' at Twyford Down, at Fairmile, at Kingston, at Newbury and so on as part of the roads protest movement, while also spearheading a twelve-year campaign of non-violent direct action to ensure that Stonehenge was open to all for the solstices and equinoxes, a campaign which he and the people he stood with can be said to have won quite comprehensively.

There is a strand of Arthuriana which runs through the cultural history of Epping Forest. Mary Wroth would have read the romances of Chrétien de Troyes, plus Spenser, of course. Tennyson rewrote the stories as love poems for the Victorians. T. E. Lawrence carried Malory's *Morte d'Arthur* in his bag throughout his campaign in Arabia. Wally Hope and William Morris both had their own obsessions with the legend. There's even a theory that Ambresbury Banks is named after Ambrosius Aurelianus, a Romano-British leader who fought the Saxons and was claimed as Arthur's uncle.

There's nothing so very remarkable to any of this. The

truth is that stories of Arthur and the Knights of the Round Table have been imprinted into our culture ever since Geoffrey of Monmouth worked them into his history of these isles. I wasn't immune myself. Susanna invited me to come to the Winter Solstice at Stonehenge on the morning of 22 December and I fully intended to, hoping to meet Arthur, maybe to be knighted by him. But when I was pulling on my thermal long johns at two o'clock in the morning ready to drive down there, my son came staggering out of his room running an impressive temperature, and by the time I'd dosed him with Calpol, settled him down and got him back to sleep it was too late to set out and I went back to bed.

Now, thinking about Mick, I remembered what Susanna had said to me about Arthur. It didn't matter, she claimed, whether Arthur was *really* a reincarnation of the real King Arthur. As I didn't believe in reincarnation or a real King Arthur this was a relief to me and I smiled and nodded enthusiastically. His importance was, she said, that by acting out the role or archetype of King Arthur, he enabled those around him to become *heroes*, heroes in the grandest, larger-than-life sense. He became an enabler for the protesters by helping people to place themselves in a narrative that was bigger, more vivid, more epic than the circumscribed reality of their everyday lives. And from this narrative they drew their strength.

Old Mick, like King Arthur, filled a role in order to allow those around him to become bigger than themselves. He was, after all, a real-life outlaw. They had come to Claremont Road to lose themselves, to go wild, and he was already hiding there. He taught them not to fear boundaries but instead to ignore them, and, in the process, almost by necessity, they idolised or rather iconicised him a little, turning him into a jewel thief, a film star, a Houdini, a hunger striker, a Robin Hood, someone who could *blow the bloody doors off* because that was what they needed him to be. And if he played along with it all, added to it all, told them ever more extravagant stories of his past, maybe he did it because he believed that was what they were all hungry for—something to nourish them, something to keep them going, a story, fiction or non-fiction, which they could tell themselves in the most difficult moments, when they were cold or scared or hungry, when vertigo welled up in them or their parents asked what the hell they thought they were doing. To put it another way, maybe the truth or otherwise of what Mick said wasn't what was most important about what Mick said. And for a while I was satisfied with that theory, even though, deep down, I knew it wouldn't hold.

I met up with Big John in a pub in Walthamstow. We drank coffee and pints of Guinness and talked about the problems he was having selling his house. He wanted to buy a camper van, take his dog and get back out on the road. Unsurprisingly considering his nickname, John is big. He's getting on a bit now and suffering with a bad hip, which meant he was walking with a stick, but his fists are great lumps of granite and when he shakes your hand it stays shook. From the top of his head down to his nose he bears a striking resemblance to Ray Winstone, and beneath that to Desperate Dan. Judging by his stories, he has spent large parts of his life in a close personal relationship with sawn-off shotguns. His dog, called The Boss, is a cross of breeds each designed to wake me up screaming in fear by night.

Susanna put me in touch with John as one of Mick's oldest friends. I knew John had done time inside and presumed that he knew Mick from all the way back then. In fact, he met him in the 1980s, when they were both involved in a variety of nefarious activities. He told me a story about Mick jumping across a gap to the balcony of a sixteenth-floor window in order to steal a stolen Pellegrini portrait, so maybe there was some substance to the old cat burglar tales. John was in prison when things kicked off at Claremont Road, but when he got out he went along, got involved and stayed involved, moving from there on to Newbury and so on.

Big John was happy, to some extent, to play Little John to Mick's Robin Hood, though Mick was an odd casting choice in many ways, John comparing him to 1970s sitcom character Harold Steptoe. He loved telling me Mick's eccentric expressions, for instance 'stone raving' as a shorthand for madness. When someone with a rather cut-glass accent was debating *progress* with him round a campfire, John remembers Mick replying, 'I don't want to be part of that progress because there's a lot of *boogaloo* attached to that.' And, whenever there was a close shave with the police, particularly while driving one of his unlicensed, untaxed, uninsured, unroadworthy vehicles, he would turn to John and smile and say, 'Two old ladies and a Bible!'

In John's recollection, Mick is less Robin Hood, more a character in an Ealing Comedy, an apposite reference point when you consider that *Passport to Pimlico* concerned the founding of a free microstate within London and *The Titfield Thunderbolt* showed the people of an English village uniting to save their train line. You could even point to *The Happy Family*, later known as *Mr Lord Says No*, which, while not an Ealing production, came out at around the same time, partakes of a similar aesthetic and tells the story of the Lord family's ultimately successful attempt to save their home from being bulldozed to make way for the Festival of Britain site. 'He had a wicked sense of humour,' recalled John. 'Although he could be a miserable bastard. Oh, he could be a mizzog,' he went on, laughing. 'The slightest thing could upset him.' Maybe it makes most sense to see the story of Claremont Road and the No M11 campaign through the lens of those Ealing Comedies, and Old Mick as a cross between a very English devil and a very dodgy Father Christmas engaged in a kind of *Only Fools and Horses* with environmental and anti-authoritarian politics. 'Old Mick would turn up and energise the whole site just with his presence,' John said. 'He had that ability. With his balding head, unshaven, he'd get out and suddenly the whole thing would come together.'

The most striking claim John made for his old friend was to do with character, with how an individual reacts to adversity. 'Front line is your true line,' he said. 'All the honest people are there.' The front line being something I've never been anywhere near, this made me a little uncomfortable, but not so much that I'd disagree and potentially move myself closer to it. 'He had an honesty about him,' John went on. 'I loved that part of him. That made me stronger…He'd never take anything from anyone…I don't know anybody that would have a bad word or could say they looked for help and he didn't help. And he never asked for anything in return or took anything in return. And I loved him. I loved him dearly.' Big John's big thumb wiped a big tear from his eye.

That was when I began to feel I was holding Mick to impossible standards. Why on earth would I expect him to tell people exactly what he'd done on the landing of a luxury house in west London all the way back in 1964? We have all, *all of us*, done things we are ashamed of during our lives. I defy you to deny it. I know I can't. And we all, *all of us*, tell lies or half-truths about those things we've done, or avoid talking about them at all costs. I know I do. And there was plenty of opportunity to see Mick's stories as half true at worst, if that's how you wanted to view it.

It *is* true, for instance, and confirmed by police reports, that Mick was living with a married woman whose husband was already inside when he was arrested in 1964, and that's part of the basic plot of *Poor Cow*. And while Nell Dunn was keen to emphasise that *Poor Cow* was a book about a woman and not the men she spends time with, she stopped short of denying that she may have come across Mick and Rose during her research. Similarly, while Mick was never a fully fledged member of the Richardson gang—and thank God for that— Nikki remembered him hanging round their scrapyard and, once again, the police files show him working, in 1960, as a scrap dealer. As for excarceration, the files show he did escape at least once, from Ashford Remand Centre, and who knows how that story could have got distorted and exaggerated over the years, not just by him but by those who retold it. And, as we've seen, Big John had a certain amount of professional respect for Mick's work as a thief: 'He was totally invisible.'

More than this, though, if you see the incident in Orme Square as the central crisis of Mick's life, then he made a considerable effort to atone for the crime he committed. According to John, the reason Mick got involved in the roads protests was a missive received by another *elderly lady*, Dolly Watson, who had lived her whole life on Claremont Road and carved her sweetheart's initials into the London plane tree out the front when she was still a kid. One day, Dolly called Russell and Mick over to look at the letter she'd been sent. It was a final eviction notice from the Department of Transport. What should she do? She really didn't want to go, she said. Mick had a look, his jaw working, that anger building up inside him, then said, 'Write back and tell 'em—if they want it, they better come and get it.'

Even as I constructed this last theory of Mick Roberts, I began to take it down. Those words, 'he'd never take anything from anyone', uttered with complete conviction and considerable emotion, perhaps struck me harder in the pub than they did when I was looking through the notes of our meeting and thinking about the times when Mick *had*, quite blatantly, taken all sorts of stuff from all sorts of people. Climbing in through a window and making off with it. And so, while I was satisfied with my theory on one level, I also knew it was a crock of horseshit on another. I fled back out to the trees, taking the train to Chingford, striding out across the Plain, working my way into the woods, following a path that led me to Three Planks Ride, cutting off next to the woodpile and back to my chosen perch, my own mighty tree. I began to think I understood it all, the liberation, the change in perspective, the sense, having lifted myself up, of being lifted up. It felt almost like a church thing, at least put like that. I imagined waiting until next summer and sitting here, my legs dangling, reading a book. I imagined drinking homemade lemonade. I imagined doing bird calls and befriending a robin. I imagined I was more comfortable than I actually was. And I wasn't comfortable at all. After a while, sitting up there really began to hurt.

My thoughts returned to Mick's funeral. Although Nikki hadn't gone, her children attended and were amazed by the turnout and the strength of feeling. John was proud that all the Traveller tribes of Britain had sent representatives, and he almost shivered as he remembered the sound of the women ululating, the flute and the harp. He laughed at the appropriateness of Mick leaving this world in a place called Covert Woods. Russell had been one of the key players in making the whole thing happen and I had heard of Ealing-style goings-on in order to liberate the body from the morgue, though all he would say on this subject was a general observation that 'we knew not to ask permission'.

The key point for Russell, though, was that Mick had requested this as a last wish. 'We only ended up burning him,' he explained, 'because he didn't want someone to own his body.' And it was this last comment, almost thrown away, that poleaxed me. It summed up my arrogance and stupidity, in that it summed up Mick's resistance to being summed up. I couldn't *own* Mick. I couldn't enclose him in some way, I couldn't use the landscape of his life as a place to grow my own cash crops and then put them up for sale. This was a man who had spent a good deal of his adult life refusing to be classified or systematised, avoiding it, twisting out of it. Making himself invisible, or else an old lady with a Bible or an outlaw or a madman or an artist. He was always beyond me, out of reach, and deliberately so. And, now that I thought about it, the same was true of all the other people I had written about, living and dead, male and female: Penny Rimbaud, Mary Wroth, Ken Campbell, Margaret Epstein, John Clare, Wally Hope, T. E. Lawrence, Gee Vaucher, Vyvyan Richards, Dick Turpin, Millican Dalton—any of them. My claims to know them were ridiculous, infantile, a kind of biographic primitive accumulation: 'To find and to see … was to use and exploit.' It was, I ended up feeling, a lie I was telling, as outlandish as the most speculative fiction. I thought back to my forest principle of mutability and realised at last that this one idea ate all the rest. There were no rules if you had a rule that no rule would ever hold. To be in the forest was to admit that everything was changeable, that nothing stayed still, the light through leaves moved by a breeze, the leaves on the floor stamped beneath feet, the fallen beech ripped from the Bagshot Bed, the path which always leads you somewhere but rarely where you want to go.

The truth is, our minds are much more like forests than they are fields, dark and labyrinthine, with sudden clearings where the light is so bright you can hardly see, then shadow and dead leaves and green moss and the husks of nuts and seeds scattered beneath the trees, the sound of wind through branches. We don't *find* ourselves in a dark forest in the middle of our lives, we realise we've always been there. We see ourselves—finally, after all these years!—and what we see is neither tidy nor rational, neither orderly nor static. It's what we do about it that matters. We can decide to run from it, marching out following a straight line. We can attempt to raze it to the ground and replant it in the hope that we will make it more productive. Or we can try to get to know the terrain better, appreciate its oddities, its uniqueness, try to get a feel for our own strange and boring, beautiful and ugly, our enclosed and unenclosable life.

For a moment I almost believed it.

The Last Tree

E ventually, much later than expected, I went back to the forest to spend the night up my tree. It took me a long time, a ludicrously long time, all things considered, and the reason for the delay was exactly the same as the reason I wanted to do it. I was afraid.

I knew this fear had no relation to fact. There are no animals in Epping Forest that could harm a fully grown man in any meaningful way. Although other people probably take to those woods by night, the chances of bumping into any of them out among those six thousand acres are tiny. And even if I were to stumble across another human being, the probability that he or she would turn out to be hostile was infinitesimal. It was never going to happen. I repeat—never. The biggest risk was probably being caught by a Forest Keeper and told off for some form of trespass, though as there are only thirteen of these tree police, that was pretty unlikely too. Strictly speaking, *camping* isn't an offence, though you're meant to get permission. And of course you're not supposed to climb the trees. None of this made it any less scary.

I have an admission to make, terrifying in itself, though in a different way. As a child, I was obsessed with A. A. Milne's stories of Winnie-the-Pooh. I'm sure that's not unusual in a middle-class Englishman of a certain age. My sister and I had a cassette player we used in the car, we would listen to story tapes on it during long journeys, and my favourite for many years was Bernard Cribbins reading *Winnie-the-Pooh* and *The House at Pooh Corner*. I know Alan Bennett's version is more fashionable these days, but in my mind there is no finer interpreter of the Bard of the Hundred Acre Wood than Cribbins. Much later I read the stories again, this time to my own children, and I stand by them. They are perfectly written in a complex, clause-heavy, Edwardian kind of way—they are very funny, and in them is a *smackerel* of something profound. But this is the shameful bit, the part which, when I think of it, makes me cringe. When I was fourteen or fifteen I wrote a full-length stage sequel to the Winnie-the-Pooh stories as a kind of left-wing allegory. And then I tried to make my friends and associates perform it. Looking back now, I have no idea what I was thinking. The idea of it is so sphincter-tightening I'm surprised I haven't turned inside out. I can only blame one of those popular philosophy/self-help type books which are everywhere now but seemed quite original at the time. It was called *The Tao of Pooh* and was basically *Zen and the Art of Motorcycle Maintenance* for repressed English people. But, probably from that book rather than any original thought I may have had while I was flouncing around casting people as Pooh and Piglet, ever since I've been stuck with the idea that the characters in *Winnie-the-Pooh* are somehow, in that vague, not-really-fully-defined way I tend to use the word, *archetypal*.

Thinking about sleeping out in the forest, a story from *The House at Pooh Corner* floated unbidden, unwanted, into my head. It's called 'Tigger is Unbounced' and in it fussy, bossy, peevish Rabbit tries to unbounce the loud, irrepressible Tigger by taking him out into the forest and getting him lost for the night. This is an act of great violence against Tigger, an attempt to destroy his essence, his self, his soul. The twist, though, is that Tigger bounces home in time for his tea and Rabbit himself gets lost, the unbouncer unbounced, shrunk down and cowed so that he returns to his burrow the next morning 'a very Small and Sorry Rabbit', and consequently, for a while, leaves everyone else alone.

As a child I loved Tigger and despised Rabbit (as of course you were supposed to), but looking at myself now I saw much more Rabbit about me than Tigger. I wondered if I could somehow crush my own Rabbit, make him 'Small and Sorry' by subjecting him to a night on his own in the forest. The fear would, finally, serve a purpose. I didn't need to kill this Rabbit, just terrify it so thoroughly it would leave me alone for a while. I wondered, in effect, if I could be transfigured, if I could use fear to destroy fear and hence act out some kind of Aristotmilnian tragedy of transformation.

The weather, however, remained bad. Not calamitously so, but cold and damp enough that I could keep putting it off, waiting for the change, for the days to get longer and hence the darkness shorter, waiting for an undefined night-time temperature I would be satisfied with, checking the wind, the five-day forecast, pretending to look for a window while looking for excuses. The longer I put it off, the more time there was for my fear to build and overwhelm me. Fear was everywhere. There was a General Election approaching, and listening to a radio show I heard a woman worried about immigration say, 'It's just about fear. We're fearful for what's going to happen to the NHS, we're fearful for the housing, we're fearful that we're starting to build this massive divide between cultures and integration and migration and it is becoming a massive crisis,' and the man she was talking to, who wasn't supposed to be bothered about immigration, replying, 'And we're the same with the fear for us. But the fear coming from both groups is the media. We don't get the truth.' After the election it was even worse, because everyone who lost said it was due to a politics of fear and everyone who won said it was about trust. No one mentioned the ownership of land or the demonic, neo-liberalising power of the Corporation of the City of London or Cartesian rationalism or the Enclosure of imagination or the burning of witches, for that matter. I began to wonder whether, through some previously unsuspected sensitivity, I'd been channelling this growing fear all along, or whether fear was merely the excuse that every loser uses.

Time kept trundling forward. All around me people reached forks in the paths of their lives, made choices, moved one way or another. My daughter passed her exams, picked her sixth-form college and what to study. A friend announced his partner was having a baby. My sister got a new flat. Another friend moved inexorably closer to finalising his divorce, held down a challenging job, looked after his kids. My uncle fell over on the bus and then moved into an old people's home. Summer came and went, and still I didn't go. While it was warm the weather no longer functioned as an excuse, but the impetus had gone out of me. There was always a reason why tonight was not the night. We were off on holiday soon, and then we'd just got back from holiday. I was supposed to be going to the pub later, or driving my son to training. Eventually so much time passed that the weather began to operate as an excuse all over again, the dark nights got longer and the days began to shrink. At some point I stopped going out to the woods altogether, as if, somehow, I had solved all my problems without even noticing.

The real problem, the deepest problem, was that I didn't believe it would make any difference. I never expected that any greater understanding would come of it, knew I wouldn't return transmogrified. Which is another way of saying I had no faith in my own conclusions, my self-help babble about exploring my own forest, which seemed as phoney and lame as any other easy answer. I didn't think I had proved anything or demonstrated any truth through my extensive researches, that the patterns I had revealed or made were anything more than pretty. I was going through the motions without even going through the motions. I didn't, deep down, believe in transformation. That was for fairy tales and ancient myths, for execrable books in the Mind, Body, Spirit section of an almost bankrupt bookshop. I believed in more of the same. I believed in bankrupt bookshops.

Eventually, though, the evening came when I gave up giving up. Or gave giving up a holiday, anyway. I had told enough people of my supposed *plan* that never to do it would be an ongoing humiliation, another question to avoid for the rest of my life, and I felt I had enough of them already. I should get it over with, I thought, then continue as before, perhaps with the addition of an amusing anecdote. I packed a bag with a torch, a blanket, a woolly hat, gloves, my hip flask of whisky and a plastic tub with some sandwiches in. I put on my long johns and doubled up my t-shirts, layering myself fatter, from

Laurel to Hardy, as if clothes could protect me from what was in my head. I waited as late as I could, hoping for some pressing reason to stay, to open a bottle of wine and watch TV. My son was already needling me, taking the piss for my failure to do what I said I'd planned, another deficit of machismo to subtract from my sheet. I felt his forehead and hoped it would be hot. My wife, knowing me best, pretended she was supportive, as the only thing which would make me do it was the friction of someone telling me I wouldn't. I managed to convince myself this was a greater affront, which was stupid of me—if I hadn't I could've stayed.

I drove up there, the light moving so much quicker than me, so that as I sat in traffic I watched the windows blacken and became scared I'd be too scared to find my tree, let alone stay. It was rush hour, that time of contradiction. The cars around me seemed angry and sluggish all at once, like wasps drowning in jam. Day and night joined in, playing tricks on me and indifferent to me without prejudice. It wasn't so very dark once I parked my car on a suburban street, once I was walking rather than sitting, moving instead of stewing, getting nearer. Out on the Plain I could smell the damp, distant mulchiness of the woods, feel diesel particulates replaced by decaying vegetation in the crucible of my lungs. I looked suspicious, of course, striding off towards the black line of trees with my rucksack and hiking boots, and while I would be lying if I said I didn't care, I didn't care as much as usual. I wanted to get there now, not to have to try to notice, no longer to have a choice. The last and latest of the dog walkers loomed past me and away, the animal panting in his wake, and if I kept checking behind me to see if I was followed, it was only because I couldn't stop.

And then I was in the trees, the quality of the sound changing, the cars instantly further away, the low hum of air moving through foliage, a startled squirrel's dash for safety resetting my heart to its beat, my steps both soft and loud in the bubble surrounding me. There was no one on the path, wide as a road, and it looked grey and forbidding as the light dipped, a timber track to an abandoned city. I tried to figure out what I was feeling and decided it was excitement. I was lying to myself, of course.

Leaving the main path, I picked my way past stumps, a long-abandoned fire, the remains of the foresters' sawings, clumps of knotty grass, the occasional puddled remnant of previous rain. My tree loomed up in front of me, just as good as I remembered, a reassuring presence, and I began to hope for the truth of the spell I'd sold myself, that somehow this bulk of living wood could protect me. I stood for a moment by its roots and looked back to the path to check if I was being watched, the ground bouncy and insubstantial beneath my feet, the forest shivering with stillness. Then I took a deep breath and pulled myself up. Hooked my bag over a lopped branch, sat and settled my back against the weight of it, feeling its solidity like arms around me.

And it worked. The magic worked. I had my epiphany. Not straight away. These things don't come instantly. They wait until you're not expecting them. I spent the first hour trying to get comfortable. After that I decided I would never be comfortable. Fear distracted me to some extent. I began to believe that a wild, bearded man was stalking me. I have no idea why he would need a beard. Also that there might be a squirrel nest above me. I seemed to recall having seen a video of a squirrel attack on YouTube and the recollection didn't reassure me.

Then the discomfort returned, only doubled or trebled. Pins and needles made phantom limbs of my actual limbs. The cold began to penetrate my layers and damp soaked up into me as if I were a discarded tissue in a dirty puddle. There were sounds in the branches above me, on the floor below me, sounds all around. Did no one and nothing sleep out here? Wind in the trees, leaves on the ground, the patter of tiny devilish feet, a herd of muntjac waiting for me with their sharp horns and nasty teeth. I say it was an hour, but it may only have been ten minutes. I didn't dare take my phone out to check, imagining the light acting as a beacon to every psychopath within a mile. All none of them.

And then it started: a switch clicked, a sudden shift occurred in my brain for no good reason, so that one minute it was fear and self-loathing and something digging into my hip and the next it was satori, enlightenment, my very own *just happened*. A quote I had come across which I'd thought about using and then rejected came bouncing back to me unbidden. I tossed it around in my head right then and marvelled at everything I'd previously missed. It was shorter and simpler in my recollection, but its original version goes like this: 'The right way to wholeness is made up, unfortunately, of fateful detours and wrong turnings. It is a *longissima via*, not straight but snakelike, a path that unites the opposites in the manner of the guiding caduceus, a path whose labyrinthine twists and turns are not lacking in terrors.'

When I first read it, what appealed to me was the image of the winding path, but now I was focused solely on that reference to *wholeness*, something I had ignored before, as either impossible or nonsensical or just a bit trite. Out in the middle of the dark forest I saw, quite abruptly, that the solution to my own trivial crisis was exactly this *wholeness*, a kind of alchemical wedding, an acceptance of my contradictory, divided, confused and confusing self. There was no solution to be had in dismemberment, in declaring war on part of my character in anger and self-recrimination. My fear was not something extraneous to me, a bird perched on a branch. It was the branch. Even if I tried to lop it off it would only grow again, more twisted perhaps, and at some strange, distorted angle.

Wholeness means embracing inconsistency and mutability, accepting your contradictions, your opposing natures, not rejecting one for another. Heat them all up in the retort of a beech tree, melt them, then bind them together. I didn't need to be purged at all, only to say yes to my strange and sometimes painful life. If I were to be an artist, did I want to be the sort who kills part of himself, who self-harms to achieve the kind of straight-line *bravery* which can only get him more lost? I remembered what Penny Rimbaud had said, about being 'so framed within conceit that I was unable to break through'. I thought about Mary Wroth and the infinite alchemical process, how I would never be done, never finished, never complete, that the act of becoming whole lay in the process of trying to become whole, in forgiving yourself, over and over again. That being the case, the sooner I started the better.

Before I could talk myself out of it, before my own Rabbit tried to persuade me to kill my own Rabbit, before I became scared about looking ridiculous, or too *hippie*, or criminally dishonest, or simply a coward, before conviction crumbled, I embraced my own fear. I held it close and acknowledged it as me. I welcomed it in, not as character fault or weakness, but as something if not beautiful then real, a door. With no time to waste, I put on my rucksack, leant forward, grasped the branch down by my feet, swung through darkness and staggered down onto black ground, my right hand stinging where the bark scraped at it. Nose and ears twitching, eyes the size of saucers, I paused for a moment, listening. Then, quick as I dared, my heart squeezing between my ribs, my mouth dry, my breathing fast and feathery, sweat all over me despite the cold, I made my dash back from the woods.

'Personally, I have faith that the Romantic
is still with us, slumbering perhaps,
waiting for its next opportunity.
And it will have opportunities.
We can count on the malignant dwarves
that run things to provide those'

CURTIS WHITE

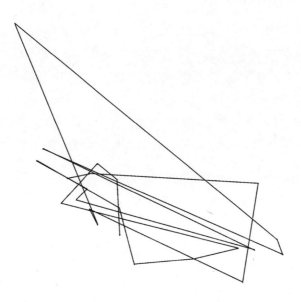

Notes and references

Chapter 1: The Skull Tree / Baldwins Hill

p. 4   'all he did was': Stephen Gardiner, *Epstein: Artist Against the Establishment*, p. 9.

p. 5   'the bad air befogs': *Epstein: Artist Against the Establishment*, p. 112.

p. 9   'In the middle': Dante Alighieri, *The Divine Comedy*, canto 1, lines 1–2.

p. 14   'a cruel, bad-tempered': Helen Garman, Kathleen's younger sister, in *Epstein: Artist Against the Establishment*, p. 219.

p. 14   'Kathleen lay wounded': Jane Babson, *The Epsteins: A Family Album*, p. 32.

p. 15   'I must protect': *Epstein: Artist Against the Establishment*, p. 227.

p. 16   'a compact little Welshman': Arthur Ransome, *The Autobiography of Arthur Ransome* (Jonathan Cape, 1976), p. 83. He goes on to describe him as 'kind, quiet, and watchful'.

p. 16   'peasant lass': ibid.

p. 16   'Peggy visited Paris': ibid.

p. 16   'Red of the Reds': June Rose, *Daemons and Angels: A Life of Jacob Epstein*, p. 234.

p. 16   'at first I was afraid': *Daemons and Angels*, p. 212.

p. 16   'looked just like a': *The Epsteins: A Family Album*, p. 27.

p. 16   'carved like a monumental': *Daemons and Angels*, p. 234.

p. 17   'a largely built Scotchwoman': *Daemons and Angels*, p. 129.

p. 17   'the dreadful ailment': *Epstein: Artist Against the Establishment*, p. 194.

p. 17   'Peggy sat like a menacing Buddha': *Daemons and Angels*, p. 196.

p. 17   'I came across a photograph': the photo is in Raquel Gilboa, ... *And there was Sculpture: Jacob Epstein's Formative Years*

1880–1930 and also on p. 6 of her essay on Whitman and Epstein: https://www.academia.edu/858512/Walt_Whitman_s_Comradeship_Epstein_s_Drawings_of_the_Calamus_Lovers_

p. 18   'Put a cloak over': *Daemons and Angels*, p. 130.

p. 21   'Sara Maitland notes': Sara Maitland, *Gossip from the Forest: The Tangled Roots of Our Forests and Fairytales*, p. 16.

p. 23   'I would go out': Jacob Epstein, *Let There Be Sculpture*, p. 165.

p. 24   'THE HYDE PARK ATROCITY' and 'The inevitable has happened': *Epstein: Artist Against the Establishment*, p. 251.

p. 25   'MONGOLIAN MORON': *Epstein: Artist Against the Establishment*, p. 319.

p. 25   'The intensity of revulsion': *Epstein: Artist Against the Establishment*, p. 245.

p. 25   'unutterably loathsome': *Epstein: Artist Against the Establishment*, p. 112.

p. 25   'some degraded Chaldean': *Epstein: Artist Against the Establishment*, p. 209.

p. 25   'some nice doctor': *Daemons and Angels*, p. 197.

p. 26   'human heads': *Epstein: Artist Against the Establishment*, p. 416.

p. 27   'I can only regret': *Daemons and Angels*, p. 188.

p. 28   'a whole TV series devoted to the phenomenon': *Dogging Tales* directed by Leo Maguire, Channel 4, 2013.

p. 30   'curiously enough': from a letter to Sylvia Epstein dated 14 December 1941, in *The Epsteins: A Family Album*, p. 55.

Chapter 2: Walthamstow Marshes / Gants Hill Library

p. 36   'The outsider's outsider': Michael Coveney, *Ken Campbell: The Great Caper*, p. 233.

p. 36   'Werner Erhard': see Suzanna Snider, 'est, Werner Erhard, and the Corporatization of Self-Help' in *The Believer*, May 2003. http://www.believermag.com/issues/200305/?read=article_snider

p. 37   'John Sessions': *Ken Campbell: The Great Caper*, p. 167

p. 38   'I'd seen the Truth': Ken Campbell, *The Bald Trilogy*, p. 292.

p. 38   'They call me a genius': *Ken Campbell: The Great Caper*, p. 233.

p. 38   'Instead of being the safe': *Ken Campbell: The Great Caper*, p. 49.

p. 38   'In them lived': Robert Pogue Harrison, *Forests: The Shadow of Civilization*, p. 61.

p. 39   'outside' and 'to keep out': *Forests: The Shadow of Civilization*, p. 69.

p. 40   'he loved the stags': *Forests: The Shadow of Civilization*, p. 75.

p. 43   'Don't *believe* in anything': *The Bald Trilogy*, p. 237.

p. 46   'When we bought this house': p. 2, http://www.douglasallen.co.uk/_assets/pdfs/52314940.pdf

p. 47   'dead ex-girlfriend' and 'the ghost of Laurence Olivier': 'Ken Campbell's Meaning of Life – A Letter to Robert Anton Wilson', https://www.youtube.com/watch?v=r_c4mw-qThk and https://www.youtube.com/watch?v=cFAIg0WfLIg

p. 47   'Derek Acorah': East London & West Essex *Guardian*, 11 January 2004, http://www.guardian-series.co.uk/news/448296.tv_crew_stumbles_upon_forest_ghost/

p. 47   'I don't think anything which isn't funny': *What Did You Do In The Warp, Daddy?*, *Arena*, BBC, 1979, at 12 minutes 40 seconds, https://www.youtube.com/watch?v=ZyZL46LeMaw

p. 51   'Now shove it up your arse': *Ken Campbell: The Great Caper*, p. 143. I misremembered the quote. It's actually 'Well, stick it up your arse and do it again', but I prefer my mistake.

p. 52   'shaman-clown': Stewart Lee, 'A Journey To Avebury', *The Quietus*, 27 June 2014, http://thequietus.com/articles/15594-julian-cope-stewart-lee-interview

p. 52   'He compared him to the first great fool': *Ken Campbell: The Great Caper,*' p. 242.

p. 52   '*reality instructors*': Mark Edmundson, 'Playing the Fool', Bookend, *New York Times*, 2 April 2000, https://www.nytimes.com/books/00/04/02/bookend/bookend.html

p. 52   'Saul Bellow': '…Simkins who loved to pity and to poke fun at the same time. He was a Reality-Instructor. Many such. I bring them out. Himmelstein is another, but cruel. It's the cruelty that gets me, not the realism', Saul Bellow, *Herzog* (Penguin, 1964), p. 36.

p. 53   'I've come to the conclusion': quoted on Frogweb: Ken Campbell, apparently a reprint of an interview with Ken from 1991 quoting an earlier interview with Robert Anton Wilson from 1976. Not exactly a rock-solid source but too good to leave out, https://web.archive.org/web/20060712180409/http://www.frogboy.freeuk.com/ken.html

p. 54 'If the regular length of a shot': Geoff Dyer, *Zona* (Canongate, 2012), pp. 9–10.

p. 54 'I've just had one of the most amazing experiences': *What Did You Do In The Warp, Daddy?*, at 3 minutes 13 seconds, https://www.youtube.com/watch?v=ZyZL46LeMaw

p. 55 'Enantiodromic Approach to Drama', 'a face is comprised', 'he's into the chastisement' and 'a vacant, inept': *The Bald Trilogy*, p. 159.

p. 56 'People get upset': *Review: The Ken Campbell Roadshow*, BBC, 1971, at 13 minutes 33 seconds, https://www.youtube.com/watch?v=OixVHxhs8vo

Chapter 3: The Doom Tree / The Lovely Illuminati Tree

p. 63 'Sir William Addison's 1945 book': *Epping Forest: Its Literary and Historical Associations* (J. M. Dent, 1945).

p. 64 'Her father, a fantastical thing': Mary Wroth, *The Countess of Montgomery's Urania* (abridged and edited by Mary Ellen Lamb), p. 169.

p. 64 'Hermaphrodite' and 'leave idle books alone': Josephine A. Roberts, *The Poems of Lady Mary Wroth*, pp. 32–3

p. 64 'redeem the time': *The Poems of Lady Mary Wroth*, p. 34.

p. 65 'Hermaphrodite': ibid.

p. 65 Mary Wroth's '*Urania*': what I'm referring to throughout is Mary Ellen Lamb's heavily edited and abridged version with modernised English, published by ACMRS in 2011.

p. 68 'little Mal': *The Poems of Lady Mary Wroth*, p. 8.

p. 69 'a very loathsome sight': Sir Dudley Carleton, in Margaret Hannay, *Mary Sidney, Lady Wroth*, p. 127.

p. 69 'his complete translation of Homer': George Chapman, 'TO THE HAPPY STARRE, DISCOUERED in our Sydneian; Asterisme; comfort of learning, Sphere of all the vertues, the Lady VVrothe.' [from *The Whole Works of Homer … in his Iliads and Odysses* [1616], http://www.english.cam.ac.uk/wroth/othertexts2.htm#chapman

p. 69 'all her books': *The Poems of Lady Mary Wroth*, p. 23.

p. 71 'cold despaire': *The Poems of Lady Mary Wroth*, p. 29.

p. 71 'her honour not touched': *The Countess of Mongomery's Urania*, p. 163.

p. 71 'She takes great libertie': John Chamberlain, in *The Poems of Lady Mary Wroth*, p. 36.

p. 77 'as Frances Yates has shown': see in particular *Giordano Bruno and the Hermetic Tradition* (1964), *The Art of Memory* (1966), *The Rosicrucian Enlightenment* (1972) and *The Occult Philosophy in the Elizabethan Age* (1979).

p. 78 'a great Chymist': John Aubrey, in Charles Nicholl, *The Chemical Theatre*, p. 15.

p. 78 'As death': Algernon Sidney, *Sidney: Court Maxims*, H. W. Blom, E. H. Mulier, R. Janse (eds.), (Cambridge University Press, 1996), p. 20.

p. 78 'so that finally man': 'Fama Fraternitas', in Frances Yates, *The Rosicrucian Enlightenment*, p. 282.

p. 79 'for the rise of Cartesian philosophy': *The Rosicrucian Enlightenment*, p. 147.

p. 79 'an alchemical fantasia': *The Rosicrucian Enlightenment*, p. 97.

Chapter 4: North Weald / Ongar

Most of the quotations in this chapter are taken from the archive of documents to be found on Wally Dean's excellent Where's Wally? blog. If you're interested in Phil Russell's story I would highly recommend giving it a browse at http://wallyhopes.blogspot.co.uk/

p. 87 'by means of fear and pity': Charles Nicholl, *The Chemical Theatre*, p. 142.

p. 88 'to celebrate oneself' and 'saying Yes to life': Friedrich Nietzsche, Section 5 of 'What I Owe to the Ancients', in *Twilight of the Idols* (1889).

p. 88 'Tragedy… is': *The Chemical Theatre*, p. 142.

p. 88 'Both alchemy and tragedy': *The Chemical Theatre*, p. 143.

p. 90 'Operation King Arthur': http://wallyhopes.blogspot.co.uk/2015/08/operation-king-arthur-every-sun-day-for.html

p. 90 'EVERY DAY IS SUN DAY': http://wallyhopes.blogspot.co.uk/2014/12/every-day-is-sun-day-manifesto-1974.html

p. 90 'FREE STONED HENGE': http://wallyhopes.blogspot.co.uk/2011/10/uk-rock-festivalscom.html

p. 91 'astral commando': *International Times*, November 1975, credited to Wally Williams, http://wallyhopes.blogspot.co.uk/2014/04/wallys-windsor-freep.html

p. 92 'Sara Maitland strolled through these': Sara Maitland, *Gossip from the Forest: The Tangled Roots of Our Forests and Fairytales*, Chapter 4, 'June – Epping Forest'.

p. 95 'I look to the revolution': 'Everyone's Wally', http://wallyhopes. blogspot.co.uk/2014/04/everyones-wally-1974.html

p. 95 'it's an effort to reclaim': quoted in George McKay, *Senseless Acts of Beauty: Cultures of Resistance since the Sixties*, p. 16.

p. 96 'Our job': *Windsor Freep*, 30 August 1973 http://wallyhopes. blogspot.co.uk/2014/04/wallys-windsor-freep.html

p. 96 'we have won': Peter Muccini, 'If love fails they'll move', Associated Press, http://wallyhopes.blogspot.co.uk/2014/04/ if-love-fails-1974.html

p. 99 'allegedly given quantities of Largactil': Penny Rimbaud, *Shibboleth: My Revolting Life*, p.183

p. 100 '*Your sun*': postcard from Cyprus. Full text reads: 'P.T.O. is a Quick advert for Legilizing P.O.T. On wayz back to de mud and Black Beer YOUR SUN Phil', http://wallyhopes.blogspot. co.uk/2013/02/wheres-wally-worldwide.html

p. 100 'I was very worried': http://wallyhopes.blogspot.co.uk/2014/ 05/the-story-of-phiilip-russell-by-dr.html

Chapter 5: Lippits Hill Lodge / St Paul's Chapel

p. 106 'hopeless hope': John Clare, *Child Harold* part 6, in *The Later Poems of John Clare*, E. Robinson and G. Summerfield (eds.), p. 69.

p. 107 'homeless at home': John Clare, 'Journey Out of Essex' lines 228–9, in *John Clare: Selected Poetry and Prose*, Merryn and Raymond Williams (eds.), p. 183.

p. 107 'to poetical': Jonathan Bate, *John Clare*, p. 525.

p. 107 'these heavy coarse': Wyndham Lewis, 'Round the London Galleries', *The Listener*, 19 January 1950, p. 116.

p. 108 'freely improvising': Bernard Grebanier, *Then Came Each Actor* (McKay, 1975) p. 21.

p. 109 'Although the dancers': Mary Ellen Lamb, *The Popular Culture of Shakespeare, Spenser and Jonson* (Routledge, 2006) p. 73.

p. 109 'Is it not better': p. xi of the Revd Alexander Dyce's introduction to William Kemp, *Kemps Nine Daies Wonder: Performed in a Daunce from London to Norwich*.

p. 110   'the old play': *Kemps Nine Daies Wonder*, p. xiv.

p. 110   'put out some money': *Kemps Nine Daies Wonder*, p. 19.

p. 110   'Master & men': *The Later Poems of John Clare*, p. 38.

p. 110   'A man there is': *The Later Poems of John Clare*, p. 37.

p. 111   'Children are fond': John Clare, *Don Juan*, in *Poems of John Clare's Madness*, Geoffrey Grigson (ed.), p. 66.

p. 112   'as simple as': J. W. and Anne Tibble, *John Clare: A Life*, p. 180.

p. 112   'what he saw': *John Clare: A Life*, p. 145.

p. 112   'Inclosure came': John Clare, 'The Mores', in Jonathan Bate, *John Clare*, p. 316.

p. 113   'There once was lanes': John Clare, *The Village Minstrel*, stanza 107.

p. 113   'major conceptual creation': John Goodridge and Kelsey Thornton, 'John Clare: the trespasser', in *John Clare in Context*, H. Haughton, A. Phillips and G. Summerfield (eds.), p. 114.

p. 113   'Clare began to realise': Adam Phillips, 'The Exposure of John Clare' in *John Clare in Context*, p. 180.

p. 114   'Poetry is not a turning loose': T. S. Eliot, 'Tradition and the Individual Talent' in *The Sacred Wood*, (Alfred A. Knopf, 1921), p. 17.

p. 114   'Very few peasants become artists': John Berger, 'The Ideal Palace', collected in *Keeping A Rendezvous* (Granta, 1992), p. 83.

p. 114   'she felt her station': *John Clare By Himself*, Eric Robinson and David Powell (eds.), (Carcanet, 1996), p. 87.

p. 115   'midnight morrice dance': *John Clare: A Life*, p. 41.

p. 115   'I used to imagine': *John Clare: A Life*, p. 40.

p. 115   'That good old': John Clare, *The Parish: A Satire*, Eric Robinson and David Powell (eds.), (Penguin, 1986), p. 33.

p. 116   'are built by those': *The Parish: A Satire*, p. 34.

p. 116   'I love thee': John Clare, 'Childish Recollections', in *The Oxford Authors: John Clare*, Eric Robinson and David Powell (eds.), (Oxford University Press, 1984), line 1, p. 40.

p. 116   'And mimicking': John Clare, 'My Mary' in *John Clare: Selected Poetry and Prose*, p. 49.

p. 116   'nervous fears': *John Clare: A Life*, p. 152.

p. 117   'thin deathlike': *John Clare: A Life*, p. 202.

p. 117   'Every discontented': *John Clare: A Life*, p. 290.

p. 118   'that Guardian spirit': *John Clare: A Life*, p. 308.

p. 118   'these dreams': *John Clare: A Life*, p. 309.

p. 118   'Ive ran the furlongs': *The Oxford Authors: John Clare*, lines 33–6, p. 262.

p. 119   'excursionists … behaved very disgracefully': from 'St Paul's demolished, 1885' in 'History of High Beach church, Part 2', http://www.highbeachchurch.org.uk/history-of-high-beach-church/history-of-high-beach-church-part-2.php

p. 120   'the grand principle': Matthew Allen, *Cases of Insanity; with Medical, Moral, and Philosophical Observations and Essays Upon Them*, p. 51.

p. 120   'The bearer': Roy Porter, '"All madness for writing": John Clare and the asylum' in *John Clare in Context*, p. 261.

p. 120   'if a small annuity': *John Clare: A Life*, p. 343.

p. 121   'Mary thou ace': John Clare, *Child Harold* in *The Later Poems of John Clare*, p. 37.

p. 121   'I slept soundly': John Clare, 'Journey Out of Essex' in *John Clare: Selected Poetry and Prose*, p. 178.

p. 122   'My guide star': John Clare, *Child Harold* in *The Later Poems of John Clare*, p. 39.

p. 123   'The computer… has become': John Berger, 'The Ideal Palace', collected in *Keeping A Rendezvous*, p. 90.

p. 124   'contrived a scheme': Peter H. Marshall, *William Godwin* (Yale University Press, 1984), p. 23.

p. 124   'as the best': Matthew Allen, *Essay on the Classification of the Insane*, Preface, p. ix.

p. 125   'a place where': *Jane Carlyle, Newly Selected Letters*, K. J. Fielding and D. R. Sorensen (eds.), (Ashgate, 2004), p. 24.

p. 125   'servants & stupid keepers': in a letter to Dr Matthew Allen, August 1841, in *The Letters of John Clare*, J. W. and Anne Tibble (eds.), p. 294.

p. 125   'a very exalted state': *Essay on the Classification of the Insane*, p. 13.

p. 126   'I have taken': *Cases of Insanity*, Preface, p. iii.

p. 126   'moral agency': *Cases of Insanity*, p. 48.

p. 126   'the planetary influence' and 'is an appendage': *Cases of Insanity*, p. 131.

p. 126   'speculative, hopeful': Thomas Carlyle, in a letter to John A. Carlyle, 5 September 1840, in *Collected Letters Online*, vol. 12,

http://carlyleletters.dukejournals.org/cgi/content/long/12/1/
lt-18400905-TC-JAC-01

p. 127   'too full of *meaning*': Alfred Tennyson in a letter to Revd H. D. Rawnsley, 1842, in *Alfred, Lord Tennyson: A Memoir by His Son* by Hallam Tennyson (Macmillan, 1897), vol. 1, p. 216.

p. 127   'the effects produced by always': *Essay on the Classification of the Insane*, p. 98.

p. 127   'forests, like fairy stories': Sara Maitland, *Gossip from the Forest: The Tangled Roots of Our Forests and Fairytales*, p. 10.

p. 127   'the feeling of being excluded from rational society': *Essay on the Classification of the Insane*, p. 89.

p. 129   'it was the decided': Christopher Ricks, *Tennyson*, p. 164.

p. 129   'have faith and all things': in a letter from Allen to Tennyson 9 May 1842, in *The Letters of Alfred Lord Tennyson: 1821–1850*, C. Y. Lang and E. F. Shannon (eds.), (Clarendon, 1982), vol. 1, p. 203.

p. 129   'I have done': *Tennyson*, p. 165.

p. 130   'collaborative film': *By Our Selves*, dir. Andrew Kötting (2015), http://www.cgplondon.org/exhibition_thumbs.php?exhib ition_id=476&show_rand=0&show_biog=0&locale= DILSTON%20GROVE. It was actually rather good, and if I don't make that clear in the text it's only because the ritual sacrifice of Iain Sinclair was more important to the success of my project than absolute honesty.

Chapter 6: Dial House / Countesthorpe Community College

Most quotes from Penny Rimbaud come directly from the interviews I recorded. Penny hasn't checked any of them because he said that once he'd spoken to me it was up to me what I did with them.

p. 138   'a nasty, worthless': George Berger, *The Story of Crass*, p. 124.

p. 142   'He shares nothing': https://www.youtube.com/watch?v= N830WfMZYqM at 4 minutes, 47 seconds.

p. 143   'Do They Owe': https://www.youtube.com/watch?v=Furrw0 VDpWM

p. 144   'James Anderton': 'God works in mysterious ways. Given my love of God and my belief in God and Jesus Christ, I have to accept that I may well be used by God in this way.' Stephen Bates, 'God using me as spokesman, says Anderton', *Daily Telegraph*, 19 January 1987.

p. 144   'They also tricked the CIA': Penny's recollection of the events are here: http://www.vice.com/en_uk/read/penny-rambaud-on-how-crass-nearly-started and you can listen to the tape here: https://www.youtube.com/watch?v=QmfLP1IOip8

p. 145   'They were way more radical': Sophie Heawood, 'Björk: I Couldn't Just Write A Disco Song', *Guardian*, 13 March 2015, http://www.theguardian.com/music/2015/mar/12/bjork-interview-vulnicura-breakup-punk-iceland-revolution

p. 145   'this time round': Crass, 'Last of the Hippies: An Hysterical Romance'. Booklet contained in *Christ – The Album*.

p. 146   'Maybe it wasn't an accident': Gavin McInnes, 'Penny Rimbaud: The Nature of Your Anger' http://www.theheavymental.com/articles/penny-rimbaud-2/

p. 147   'The fanciful dream': Penny Rimbaud, introduction to the 2009 reissue of 'Last of the Hippies: An Hysterical Romance', p. xviii.

p. 154   'THERE IS NO AUTHORITY': 'Last of the Hippies: An Hysterical Romance', p. 98.

Chapter 7: The Green Man / Claremont Road

p. 160   'and our Ladys hearts': William Addison, *Epping Forest* (J. M. Dent, 1945), p. 140.

p. 161   'the day of the supremacy': in Joe Moran, *On Roads: A Hidden History*, p. 204.

p. 162   'I continued reading': John Tyme, *Motorways versus Democracy*, p. 17.

p. 162   'consummate evil': *Motorways versus Democracy*, p. 1.

p. 162   'a showpiece of British': *On Roads: A Hidden History*, p. 209.

p. 163   'a great car economy': *On Roads: A Hidden History*, p. 212.

p. 163   'the biggest road': ibid.

p. 165   'Frestonia': http://www.frestonia.org

p. 166   'Long may Wanstonia': *On Roads: A Hidden History*, p. 215.

p. 166   'the longest and most expensive': Sandy McCreery, 'The Claremont Road Situation' in I. Borden, J. Kerr, J. Rendell and A. Pivaro (eds.), *The Unknown City: Contesting Architecture and Social Space*, p. 230.

p. 166   'an extraordinary festival': *The Unknown City: Contesting Architecture and Social Space*, p. 231.

p. 167 'Place is lived space': *The Unknown City: Contesting Architecture and Social Space*, p. 241.

p. 167 'The No M11 campaign was': John Jordan, 'The art of necessity: the subversive imagination of anti-road protest and Reclaim the Streets' in George McKay (ed.), *DiY Culture: Party and Protest in Nineties' Britain*, p. 132.

p. 169 'We are basically about taking back': *DiY Culture: Party and Protest in Nineties' Britain*, p. 139–40.

p. 170 '*Life in the Fast Lane – the No M11 Story*': https://www.youtube.com/watch?v=49wKgtOooqs
There are parallels to be drawn between the way in which the commons were carved up among landowners for their personal gain and the way in which the internet has become increasingly 'privatised' by big corporations. Whether anti-capitalists using YouTube is a form of subversion on their part, or co-option on the part of those corporations, remains to be seen. There's a sense in which, like John Clare working in the gardens of the asylum, we are all making produce for which we're not paid.

p. 172 'There is another video on YouTube': https://www.youtube.com/watch?v=q9NuW8ogoWo

p. 175 'When I was 14 years old we were evicted': letter included in the case file for Defendant: ROBERTS, Michael Roger and others. Charge: Larceny; 16 June 1959. Held at the National Archives, Kew.

Chapter 8: Theydon Bois / Ambresbury Banks

p. 188 'He was found': *International Times*, Volume Maya, Issue 3, December 1974, http://www.internationaltimes.it/archive/index.php?year=1974&volume=Maya&issue=3&item=IT_1974-12-01_D-Maya_Iss-3_007

p. 188 'He lived well into' and 'extremely cheerful': West Essex *Gazette & Independent*, 13 December 1974, p. 1.

p. 188 'on a footpath': West Essex *Gazette & Independent*, 11 October 1974, p. 1.

p. 194 '*closet exhibitionist*': 'What's envied is the gift for theatrical self-transformation, the way they are able to loosen and make ambiguous their connection to a real life through the imposition of talent. The exhibitionism of the superior artist is connected to his imagination; fiction is for him at once playful hypothesis and serious supposition, an imaginative form of inquiry—everything that exhibitionism is not. It is, if anything, closet exhibitionism, exhibitionism in hiding.' Philip Roth, *The Counterlife* (Vintage, 2005) p. 214. The words are those of dead Nathan Zuckerman's young editor at his funeral, so probably represent the exact opposite of what Roth himself believes.

p. 198   'transgressio': J. H. Baker, *An Introduction to English Legal History*, p. 71.

p. 198   'Our earliest examples': Theodore Plucknett, *A Concise History of the Common Law*, p. 465.

p. 198   'violent invasions by marauders': *A Concise History of the Common Law*, p. 466.

p. 200   'by nineteen out of twenty': A. J. Peacock, *Bread or Blood*, p. 19.

p. 200   'There were many whose eyes': John and Barbara Hammond, *The Village Labourer 1760–1832*, p. 37.

p. 200   'Cloaks for ye Greatest Villainies': Janet Neeson, *Commoners: common right, enclosure and social change in England 1700–1820*, p. 161.

p. 200   'in sauntering after his cattle': John and Barbara Hammond, *The Village Labourer 1760–1832*, p. 38.

p. 200   'nests of sloth': Harry Hopkins, *The Long Affray: The Poaching Wars 1760–1914*, p. 23.

p. 200   'nursery and resort': Board of Agriculture (Great Britain), *General View of the Agriculture of the County of Essex* (London, 1807), vol. 2, p. 153.

p. 201   'if this is the best': Jack London, *The People of the Abyss*. My edition is a rather ugly small-format edition from Journeyman Press (1978) with tiny type and in here it's p. 115.

p. 202   'nothing but the terror': *The Village Labourer 1760–1832*, p. 164.

p. 203   'five major Bills': this line of argument comes from Peter Linebaugh in *The London Hanged*, p. 16 ff.

p. 203   'There is hardly a': E. P. Thompson, *Whigs and Hunters: The Origin of the Black Act*, p. 22.

p. 205 'naturally of a warm': Anonymous, *The History of the Blacks of Waltham in Hampshire and those under like Denomination in Berkshire*, p. 3.

p. 205 'to inform the Gentleman': *The History of the Blacks of Waltham in Hampshire*, p. 9.

p. 205 'but 15 of his Sooty Tribe': *The History of the Blacks of Waltham in Hampshire*, p. 7.

p. 206 'The Hanoverian accession': *Whigs and Hunters*, p. 99.

p. 206 'had no other design than': *The History of the Blacks of Waltham in Hampshire*, p. 7.

p. 207 '*The Forked Forest Path*': http://olafureliasson.net/archive/artwork/WEK101549/the-forked-forest-path

p. 208 '*A Line Made by Walking*': http://www.tate.org.uk/art/artworks/long-a-line-made-by-walking-ar00142

p. 208 'In all fictional works': Jorge Luis Borges, trans. Donald A. Yates, 'The Garden of Forking Paths' in *Labyrinths* (Penguin, 1970), p. 51.

p. 208 'an incomplete, but not false, image': 'The Garden of Forking Paths', p. 53.

Chapter 9: Hollow Ponds / Highams Park

p. 215 'There indeed robbery was organised': Thomas Babington Macaulay, *The History of England, from the Accession of James II*, vol. 5, p. 87.

p. 216 'King Orronoko': William Addison, *Epping Forest: Its Literary and Historical Associations*, p. 128.

p. 217 'Upon such a view': Henry Fielding, *An Enquiry Into the Causes of the Late Increase of Robbers and Related Writings*, Malvin R. Zirker (ed.), (Clarendon, 1988), p. 131.

p. 217 'educational stodge palace': the William Morris Gallery has since been extensively refitted so that it is now much brighter and more welcoming. In 2013 it even won the Museum of the Year award. Something about it, though, still leaves me feeling very tired.

p. 218 'the boundary' and 'an asylum': Maurice Keen, *The Outlaws of Medieval Legend*, p. 2.

p. 219 'Peter Linebaugh has detailed': Peter Linebaugh, *The London Hanged*, p. 184 ff.

p. 220   'diverse disorderly': James Sharpe, *Dick Turpin: The Myth of the English Highwayman*, p. 150.

p. 220   'God damn your Blood': Richard Bayes, *The Genuine History of the Life of Richard Turpin*, p. 4. Bayes, rather prissily, leaves gaps for all the curse words but I filled in the blanks.

p. 220   'You can see why E. P. Thompson': E. P. Thompson, *Whigs and Hunters: The Origin of the Black Act*, p. 82.

p. 221   'In this Cave': *The Genuine History of the Life of Richard Turpin*, p. 15.

p. 226   'This excavation': *Epping Forest*, p. 131.

p. 227   'If he would only stay': *The Genuine History of the Life of Richard Turpin*, p. 20.

p. 229   'Nor had he to wait': William Harrison Ainsworth, *Rookwood: A Romance*, Volume III, Book 4, Chapter VIII, p. 333.

p. 230   'Is she not beautiful?': *Rookwood*, Chapter V, p. 282.

p. 230   'There was a dreadful gasp': *Rookwood*, Chapter VIII, p. 348.

p. 230   'Well do I remember': *Rookwood*, Preface to the 1853 edition, pp. xxvii–xxviii. Ainsworth seems to have added various prefaces and ever more dramatic dedications to his mother as time went by. Here is a man who knew how to enjoy his celebrity.

p. 231   'It's been argued, and persuasively': David Wootton, 'The road is still open' in *London Review of Books*, vol. 27, no. 3, 3 February 2005, http://www.lrb.co.uk/v27/n03/david-wootton/the-road-is-still-open

p. 231   'All the Chartists': Mary Russell Mitford, *The Life of Mary Russell Mitford*, Revd A. L'Estrange (ed.), (Richard Bentley, 1870), vol. 3, p. 106.

p. 232   'If ever there was a publication': *The Examiner*, no. 1691, 28 June 1840, p. 2.

p. 232   'Some of the children': *The London Hanged*, p. 7.

p. 234   'the principles of fairness': Ben Leapman, 'Waite: free police killer', *London Evening Standard*, 27 July 2004, http://www.standard.co.uk/news/waite-free-police-killer-6963588.html

p. 235   'torn limb from limb': Ian Gallagher, 'Police killer Harry Roberts's five-year terror campaign to silence woman who kept him behind bars', *Mail on Sunday*, 19 April 2009, http://www.dailymail.co.uk/news/article-1171773/Police-killer-

Harry-Robertss-year-terror-campaign-silence-woman-kept-bars.html

p. 235   'Harry Roberts is our friend': for instance, https://www.youtube.com/watch?v=7G6WIqYD8SY from 19 seconds in.

Chapter 10: Crown Hill / The Cloven Tree

p. 243   'excarceration': Peter Linebaugh, *The London Hanged*, p. 3.

p. 243   'delight in the finite' and 'in no way sought': Roland Barthes, trans. A. Lavers, 'The Nautilus and the Drunken Boat' in *Mythologies* (Jonathan Cape, 1972), p. 65.

p. 244   'Solitude has the peculiar': Adam Sharr, *Heidegger's Hut*, p. 65.

p. 244   'Why do I sleep outdoors?': Roger Deakin, *Wildwood*, p. 12.

p. 247   'Welsh-American': John E. Mack, *A Prince of Our Disorder: Life of T.E. Lawrence*, p. 60.

p. 248   'there was a craving': T. E. Lawrence, *Seven Pillars of Wisdom*, Chapter CIII, p. 580.

p. 249   'two would have been enough': Vladimir Nabokov, trans. and ed. O. Voronina and B. Boyd, *Letters to Véra* (Penguin, 2014), p. 420. To be fair, he goes on to say it is 'wonderfully written in places'.

p. 249   'four hundred and fifty pages into the complete book': or at least into my Penguin Twentieth Century Classics edition.

p. 250   'I remembered smiling idly at him': *Seven Pillars of Wisdom*, Chapter LXXX, p. 454.

p. 250   'I think Lawrence enacted': *T. E. Lawrence and Arabia*, BBC documentary, 1986, https://www.youtube.com/watch?v=kk2L9JrTbjE at 3 minutes and 24 seconds.

p. 251   'It's part of my atonement': letter to Charlotte Shaw, January 1924, quoted in Mary Helen Fernald, 'The literary relationship between T.E. Lawrence and Mr and Mrs Bernard Shaw: A Thesis' (Orono, 1962) p. 26.

p. 252   'He has invented': *A Prince of Our Disorder*, p. 277.

p. 253   'He hated the thought of sex': *T. E. Lawrence and Arabia*, BBC documentary, 1986, https://www.youtube.com/watch?v=3lVRQ3nFKsk at 1 minute and 11 seconds.

p. 253   'began indifferently to slake': *Seven Pillars of Wisdom*, Chapter I, p. 28.

p. 253   'A slight figure': Harold Orlans, *T.E. Lawrence: Biography of a Broken Hero* (McFarland, 2002), p. 165.

p. 254   'It was love at first sight': ibid.

p. 254   'most exquisitely': Vyvyan Richards, *Portrait of T.E. Lawrence*, p. 49.

p. 255   'and we must then interview builders': letter to Richards, 1 September 1919, included in Malcolm Brown (ed.), *The Letters of T.E. Lawrence*, p. 168.

p. 255   'His dream of the perfect hall': *Portrait of T.E. Lawrence*, p. 180.

p. 255   'worn nerves': *Portrait of T.E. Lawrence*, p. 181.

p. 255   'All was to be': ibid.

p. 256   'I imagine leaves': letter to Bruce Rogers, 6 May 1935.

p. 257   'The pride of [Epping] Forest': Nikolaus Pevsner and Enid Radcliffe, *The Buildings of England: Essex* (Penguin, 1979) p. 173.

p. 259   'I gave up my job': M. D. Entwistle, *Millican Dalton: A Search for Romance and Freedom*, p. 12.

p. 260   'gipsies, hawkers': Epping Forest Act 1878, https://www.cityof london.gov.uk/things-to-do/green-spaces/epping-forest/about-us/Documents/1878-Epping-Forest-Act.pdf

p. 261   'it's the ground that's yours': Italo Calvino, trans. Isabel Quigly, 'Baron in the Trees' in *Our Ancestors* (Minerva 1992), p. 92.

p. 262   'The pleasure of money is unending': George Barker, *The True Confession of George Barker* (Parton, 1957), Book II, p. 7.

p. 262   'It is quite impossible': Sigmund Freud, *On the Universal Tendency to Debasement in the Sphere of Love*. I read it here, page 6, column 2: http://math.msgsu.edu.tr/~dpierce/Texts/Freud/freud_debasement.pdf

p. 263   'the very incapacity': ibid.

p. 264   '*Return of Owners Land*': see Kevin Cahill, *Who Owns Britain*, particularly chapter 3, pp. 29–57.

p. 265   'First the dark forest': Ken Frieden, *Freud's Dream of Interpretation* (Albany, 1990) p. 11.

Chapter 11: Ongar Park / Comical Corner

Once again, all Penny Rimbaud quotes are from our interviews except:

p. 270   'Ubi Dwyer...Paul Morozzo': see pages 95 and 169, respectively.

p. 274   'If I'm going to close down': George Berger, *The Story of Crass*, p. 277.

p. 278 'Take our attitude towards ownership of land': see Kevin Cahill, *Who Owns Britain* (Canongate, 2001), particularly p. 199.

p. 279 'administered population': Curtis White, *The Middle Mind* (Penguin, 2005), 'Introduction to the British Edition', p. xiii.

p. 279 'We are told, of course': *The Middle Mind*, p. 188.

p. 282 'We are the revolution': from the original Charter of the Woodcraft Folk, quoted in Leslie Paul, *Angry Young Man*, p. 63 (Faber, 1952).

p. 282 'In this matter': René Descartes, trans. John J. Blom, *Discourse On Method* (1977), pp. 129–30.

p. 285 'So that's my interest, almost my revulsion, with Cartesian thinking': interestingly, this notion of a 'permanent point of judgement' is similar to David Hockney's objection to conventional perspective in paintings.

p. 285 'Sven Lindqvist': in particular Sven Lindqvist, *Exterminate All the Brutes* (1992).

p. 285 'the rage for order and symmetry': John and Barbara Hammond, *The Village Labourer 1760–1832*, p. 40.

p. 287 'the law of identity': Robert Pogue Harrison, *Forests: The Shadow of Civilization*, p. 63.

p. 287 'the reality that we refer to': *The Middle Mind*, p. 3.

p. 287 'The Tragedy of the Commons': by Garrett Hardin in *Science*, 13 December 1968, vol. 162, no. 3859, pp. 1243–8 http://www. sciencemag.org/content/162/3859/1243.full. As well as doing exactly what it says in its title, 'A Short History of Enclosure in Britain' by Simon Fairlie is an excellent introduction to and critique of Hardin http://www.thelandmagazine.org.uk/ articles/short-history-enclosure-britain

p. 288 'There is no such thing': *Woman's Own* magazine, 31 October 1987. Typical that she should choose Halloween to scare us.

p. 289 'whereas he was probably essentially': Gavin McInnes, 'Penny Rimbaud: The Nature of Your Anger' http://www.cvltnation. com/the-heavy-mental-interview-with-penny-rimbaud/

Chapter 12: Loughton Hall / Lopping Hall

p. 296 'an excellent building': Nikolaus Pevsner and Enid Radcliffe, *The Buildings of England: Essex* (Penguin, 1979), p. 288.

p. 297   'The bell tolled on': Charles Dickens, *Barnaby Rudge* (Chapman & Hall, 1841) p. 255.

p. 298   'The Library is still': from a photocopy given to me by the manager at Loughton Hall, typed up by Vic Barham of the Chigwell and Loughton History Society, 1984.

p. 298   'although, itt pierce': Mary Wroth, the fifth of her crown of sonnets, in *The Poems of Lady Mary Wroth*, Josephine A. Roberts (ed.), p. 130.

p. 298   'Cleere as th'ayre': Mary Wroth, the third of her crown of sonnets, in *The Poems of Lady Mary Wroth*, p. 129.

p. 298   'Night, welcome art thou': Mary Wroth, 37, in *The Poems of Lady Mary Wroth*, p. 109.

p. 298   'Leave the discourse of Venus': Mary Wroth, 103, in *The Poems of Lady Mary Wroth*, p. 142.

p. 300   'in its practical working': Lord Eversley, *Commons, Forests and Footpaths*, p. 16.

p. 300   'over one thousand square miles': incidentally, exactly the same quantity of English land as was licensed for fracking in August 2015: http://www.theguardian.com/environment/2015/aug/18/1000-sq-miles-england-opened-up-fracking-new-round-licences

p. 301   'a spendthrift, a profligate': *The Gentleman's Magazine and Historical Review*, August 1857, p. 216.

p. 302   'a considerable tradition': Sara Maitland, *Gossip from the Forest: The Tangled Roots of Our Forests and Fairytales*, p. 90.

p. 304   '*officially*, not only to': Nicholas Shaxson, *Treasure Islands: Tax Havens and the Men Who Stole the World*, p. 251.

p. 304   'Over and over again we have seen': George Monbiot, 'The medieval, unaccountable Corporation of London is ripe for protest', *Guardian*, 31 October 2011, http://www.theguardian.com/commentisfree/2011/oct/31/corporation-london-city-medieval. This is where Monbiot's 'Babylon' quote comes from too.

p. 304   'It is a very intelligent demon': *Treasure Islands*, p. 270.

p. 305   'this body, with a keen eye': *Commons, Forests and Footpaths*, p. 92.

p. 305   'London would not be London': *Treasure Islands*, p. 259.

p. 305   'instead of being a congerie': *Commons, Forests and Footpaths*, p. 93.

p. 306   'With the aid of a large': *Commons, Forests and Footpaths*, p. 103.

p. 307   'very unfairly': *Commons, Forests and Footpaths*, p. 105.

p. 308   'Thus of dead leaves': Mary Wroth, 19, in *The Poems of Lady Mary Wroth*, p. 98.

p. 309   'Wroth's unit of thought': Mary Ellen Lamb, introduction to *The Countess of Montgomery's Urania*, p. 7.

p. 309   'It's Fort's opinion': Ken Campbell, *The Bald Trilogy*, p. 39.

p. 311   'Pamphilia is the queen': *The Countess of Montgomery's Urania*, p. 180.

p. 311   'Amphilanthus was extremely': *The Countess of Montgomery's Urania*, p. 253.

p. 312   'In this strang labourinth': Mary Wroth, first line of the crown of sonnets in *The Poems of Lady Mary Wroth*, p. 127.

p. 313   'This is as strange': William Shakespeare, *The Tempest*, Act V, sc. i, lines 254–6. I mentioned Ken Campbell's appearance in Derek Jarman's film of *The Tempest* earlier but failed to point out that Heathcote Williams, who has cropped up again and again in this book, played Prospero.

p. 314   'The witch-hunt was one': Silvia Federici, *Caliban and the Witch*, p. 165.

p. 315   'as a receptacle': *Caliban and the Witch*, p. 141.

p. 315   'Magic kills': *Caliban and the Witch*, p. 142.

p. 315   'the ways in which a book can be infinite': Jorge Luis Borges, trans. Donald A. Yates, 'The Garden of Forking Paths' in *Labyrinths* (Penguin, 1970), p. 51.

p. 316   'These our actors': William Shakespeare, *The Tempest*, Act IV, sc. i, lines 148–54.

p. 318   'Alzheimer's is': David Shenk, *The Forgetting*, p. 255.

p. 318   'the nursing center encourages residents': *Dosher Memorial Hospital 75th Anniversary Magazine* at http://www.dosher.org/flash/ p. 55.

p. 318   'she… leaves behind': obituary of Peggy Jean Lewis at http://www.legacy.com/obituaries/starnewsonline/obituary.aspx?n=peggy-jean-lewis&pid=145568764

p. 319   'I think she wanted her daughters to excel': Cressida Connolly, 'Kitty Godley obituary', *Guardian*, 19 January 2011,

http://www.theguardian.com/artanddesign/2011/jan/19/
kitty-godley-obituary

p. 319   'The dementia of her final years': ibid.

Chapter 13: Orme Square / Covert Woods

All quotes referring directly to criminal cases involving Mick Roberts are taken from the statements and other documents contained in the case files at the National Archives, Kew.

p. 327   'barbaric crime': 'Sentence for barbaric crime', *The Times*, 26 January 1965.

p. 327   'I was researching Mrs Harvey's husband, Wilfred Harvey': 'Harvey was a crook and was helping himself to money. The fraud was that he acquired various companies and paid X for them but he entered them in the books as three or four times X, and the difference went to the credit of his personal account.' David Hobson, quoted in Derek Matthews and Jim Pirie, *The Auditors Talk: An Oral History of a Profession from the 1920s to the Present Day* (Garland, 2000), p. 277.

p. 328   'suffered massive financial irregularities': Tom Rosenthal, 'Obituary: Edmund Fisher', *Independent*, 4 March 1995, http://www.independent.co.uk/news/people/obituary-edmund-fisher-1609911.html

p. 329   'the method of escape': '27 Escapes from Dartmoor in three years', *The Times*, 31 March 1964.

p. 332   'His story is brilliantly told': C. J. Stone, *The Trials of Arthur*. Be sure to buy the later edition published by The Big Hand in 2012, as it contains a lot more material than the original book did.

p. 336   'Ealing Comedy': the identification of roads protests with Ealing Comedies is, if I remember rightly, drawn from Joe Moran, *On Roads: A Hidden History*.

p. 341   'To find and to see': Adam Phillips, 'The Exposure of John Clare' in *John Clare in Context*, H. Haughton, A. Phillips and G. Summerfield (eds.), p. 180.

Epilogue: The Last Tree

p. 347   'A very Small and Sorry Rabbit': A. A. Milne and E. H. Shepard, *Winnie-The-Pooh: The Complete Collection of Stories and Poems* (Egmont, 1994), p. 213.

p. 352   'The right way to wholeness': my fearfulness is shown by my refusal to name the quote's owner in the text. It's from C. G. Jung, trans. R. F. C. Hull, *Psychology and Alchemy* (Routledge & Kegan Paul, 1968), p. 6, though I first read it in Charles Nicholl, *The Chemical Theatre*, p. 143.

p. 355   'Personally, I have faith': Martin Riker, 'The "Splendidly Cranky" Utopian: An Interview with Curtis White', *Paris Review*, 3 December 2015. The 'R' word plays much the same role in this book as 'time' did for Ts'ui Pên in Borges's 'The Garden of Forking Paths', http://www.theparisreview. org/blog/2015/12/03/the-splendidly-cranky-utopian-an-interview-with-curtis-white/

Bibliography

Addison, William, *Epping Forest: Its Literary and Historical Associations* (J. M. Dent, 1945)

Addison, William, *The Old Roads of England* (Batsford, 1980)

Ainsworth, William Harrison, *Rookwood: A Romance* (London, 1834)

Allen, Matthew, *Cases of Insanity; with Medical, Moral, and Philosophical Observations and Essays Upon Them* (George Swire, 1831)

Allen, Matthew, *Essay on the Classification of the Insane* (John Taylor, 1837)

Anonymous, *The History of the Blacks of Waltham in Hampshire and those under the like Denomination in Berkshire* (A. Moore, 1723)

Babson, Jane F., *The Epsteins: A Family Album* (Taylor Hall, 1984)

Bachofen, J. J., *Myth, Religion and Mother Right: Selected Writings* (Routledge & Kegan Paul, 1967)

Baker, J. H., *An Introduction to English Legal History* (Butterworths, 1990)

Barlow, Derek, *Dick Turpin and the Gregory Gang* (Phillimore, 1973)

Bate, Jonathan, *John Clare: A Biography* (Picador, 2004)

Bayes, Richard, *The Genuine History of the Life of Richard Turpin* (J. Standen, 1739)

Beam, Alan, *Rehearsal for the Year 2000* (Revelaction Press, 1976)

Berger, George, *The Story of Crass* (Omnibus, 2006)

Borden, I., Kerr, J., Rendell, J. and Pivaro, A., *The Unknown City: Contesting Architecture and Social Space* (MIT Press, 2001)

Bradbrook, Muriel Clara, *The School of Night: A Study in the Literary Relationships of Sir Walter Raleigh* (Russell & Russell, 1965)

Brennan, Michael, *The Sidneys of Penshurst and the Monarchy, 1500–1700* (Ashgate, 2006)

Brown, Malcolm (ed.), *The Letters of T.E. Lawrence* (OUP, 1991)

Burkert, Walter (trans. John Raffan), *Greek Religion: Archaic and Classical* (Basil Blackwell, 1985)

Buxton, Edward North, *Epping Forest* (Edward Stanford, 1884)

Cahill, Kevin, *Who Owns Britain* (Canongate, 2001)

Campbell, Ken, *The Bald Trilogy* (Methuen, 1995)

Chapman, George, *The Whole Works of Homer . . . in his Iliads and Odysses* (1616)

Clare, John (eds. E. Robinson and G. Summerfield), *The Later Poems of John Clare 1837–1864* (Clarendon, 1984)

Clare, John (eds. J. W. and Anne Tibble), *The Letters of John Clare* (Routledge & Kegan Paul, 1951)

Clare, John (ed. Geoffrey Grigson), *The Poems of John Clare's Madness* (Routledge & Kegan Paul, 1949)

Cline, Ann, *A Hut of One's Own: Life Outside the Circle of Architecture* (MIT Press, 1997)

Connolly, Cressida, *The Rare and the Beautiful: The Lives of the Garmans* (Fourth Estate, 2004)

Cope, Julian, *The Modern Antiquarian* (Thorsons, 1998)

Cork, Richard, *Jacob Epstein* (Tate Gallery Publishing, 1999)

Coveney, Michael, *Ken Campbell: The Great Caper* (Nick Hern, 2011)

Cox, John Charles, *The Royal Forests of England* (Methuen, 1905)

Deakin, Roger, *Wildwood: A Journey Through Trees* (Hamish Hamilton, 2007)

Debus, Allen G., *The English Paracelsians* (Oldbourne, 1965)

Descartes, René (trans. John J. Blom), *Discourse On Method* (1637, Harper & Row, 1977)

Donaldson, Ian, (ed.), *The Oxford Authors: Ben Jonson* (OUP, 1985)

Entwistle, M. D., *Millican Dalton: A Search for Romance and Freedom* (Mountainmere Research, 2004)

Epstein, Jacob, *Let There Be Sculpture* (Michael Joseph, 1940)

Evelyn, John, *Sylva* (John Martyn and James Allestry, 1664)

Eversley, Lord, *Commons, Forests and Footpaths* (Cassell, 1910)

Fairlie, Simon, 'A Short History of Enclosure in Britain', in *The Land Magazine*, issue 7 (2009)

Federici, Silvia, *Caliban and the Witch* (Autonomedia, 2004)

Fernald, Mary Helen, *The Literary Relationship between T.E. Lawrence and Mr and Mrs Bernard Shaw* (Orono, 1962)

Fisher, W. R., *The Forest of Essex* (Butterworths, 1887)

Fletcher, Angus, *Colors of the Mind* (Harvard Univeristy Press, 1991)

Fraser, Angus, *The Gypsies* (Blackwell, 1992)

Frazer, James George, *The Golden Bough* (Macmillan, 1922)

Gardiner, Stephen, *Epstein: Artist Against the Establishment* (Michael Joseph, 1992)

Gatti, Hilary, *Essays on Giordano Bruno* (Princeton University Press, 2011)

Giddens, Anthony, *The Transformation of Intimacy* (Polity, 1992)

Gilboa, Raquel,... *And there was Sculpture: Jacob Epstein's Formative Years 1880–1930* (Paul Holberton, 2009)

Hagger, Nicholas, *A View of Epping Forest* (O Books, 2012)

Hammond, J. L. and Barbara, *The Village Labourer 1760–1832* (Longmans, 1920)

Hannay, Margaret P., *Mary Sidney, Lady Wroth* (Ashgate, 2010)

Harrison, Robert Pogue, *Forests: The Shadow of Civilization* (University of Chicago Press, 1992)

Harvey, Graham (ed.), *Shamanism: A Reader* (Routledge, 2002)

Haughton, H., Phillips, A. and Summerfield, G. (eds.), *John Clare In Context* (CUP, 1994)

Hayward, Arthur L. (ed.), *Lives of the Most Remarkable Criminals who have been Condemned and Executed for Murder, Highway Robberies, Housebreaking, Street Robberies, Coining, or Other Offences* (1735; Routledge, 2002)

Hearn, Karen, *Marcus Gheeraerts II: Elizabethan Artist In Focus* (Tate Publishing, 2002)

Holdsworth, Sir William, *A History of English Law* (1908; Sweet & Maxwell, 1991), vol. 3

Hopkins, Harry, *The Long Affray: The Poaching Wars 1760–1914* (Secker & Warburg, 1985)

Huffman, William (ed.), *Robert Fludd: Essential Readings* (Aquarian/ Thorsons, 1992)

Keen, Maurice, *The Outlaws of Medieval Legend* (Routledge & Kegan Paul, 1961)

Kemp, William, *Kemps Nine Daies Wonder* (1600; Camden Society, 1840)

Kenrick, Donald, *Gypsies: from the Ganges to the Thames* (University of Hertfordshire Press, 2004)

Knightley, Philip, and Simpson, Colin, *The Secret Lives of Lawrence of Arabia* (Nelson, 1969)

Korda, Michael, *Hero: The Life and Legend of Lawrence of Arabia* (JR, 2011)

Kyll, Thomas, *The Trial of the Notorious Highwayman Richard Turpin* (1739)

Lamb, Mary Ellen, *The Popular Culture of Shakespeare, Spenser and Jonson* (Routledge, 2006)

Lawrence, T. E., *Seven Pillars of Wisdom: A Triumph* (Jonathan Cape, 1935)

Levi, Peter, *Tennyson* (Macmillan, 1993)

Levy, Gertrude Rachel, *The Gate of Horn* (Faber & Faber, 1948)

Linebaugh, Peter, *Stop, Thief! The Commons, Enclosures, and Resistance* (PM Press, 2014)

Linebaugh, Peter, *The London Hanged: Crime and Civil Society in the Eighteenth Century* (Penguin, 1991)

Linebaugh, Peter, *The Magna Carta Manifesto* (University of California Press, 2008)

London, Jack, *People of the Abyss* (Macmillan, 1903)

MacCulloch, Diarmad, *Reformation* (Allen Lane, 2003)

Mack, John E., *A Prince of our Disorder: The Life of T.E. Lawrence* (Weidenfeld & Nicolson, 1976)

Maitland, Sara, *Gossip from the Forest: The Tangled Roots of Our Forests and Fairytales* (Granta, 2012)

McGilligan, Patrick, *Alfred Hitchcock: A Life in Darkness and Light* (Wiley, 2003)

McKay, George, *Senseless Acts of Beauty* (Verso, 1996)

McKay, George (ed.), *DiY Culture: Party and Protest in Nineties' Britain* (Verso, 1998)

Merrifield, Jeff, *Seeker! Ken Campbell: His Five Amazing Lives* (Playback, 2011)

Miles, Rosalind, *Jonson: His Life and Work* (Routledge & Kegan Paul, 1986)

Monod, P., Pittock, M., Szechi, D., (eds.), *Loyalty and Identity: Jacobites at Home and Abroad* (Palgrave Macmillan, 2010)

Moran, Joe, *On Roads: A Hidden History* (Profile, 2009)

Neeson, J. M., *Commoners: common right, enclosure and social change in England 1700–1820* (Cambridge Univeristy Press, 1993)

Nicholl, Charles, *The Chemical Theatre* (Routledge & Kegan Paul, 1980)

Nicholson, Virginia, *Among the Bohemians* (Viking, 2002)

Nord, Dorah Epstein, *Gypsies and the British Imagination, 1807–1930* (Columbia University Press, 2006)

Peacock, A. J., *Bread or Blood* (Gollancz, 1965)

Perceval, P. J. S., *London's Forest* (J. M. Dent, 1909)

Plucknett, Theodore F. T., *A Concise History of the Common Law* (Lawyers Co-operative, 1929)

Porteous, Alexander, *The Forest in Folklore and Mythology* (Allen & Unwin, 1928)

Powell, W. R. (ed.), *A History of the County of Essex: Volume IV, Ongar Hundred* (Victoria County History, 1956)

Rackham, Oliver, *Trees and Woodland in the British Landscape* (J. M. Dent, 1990)

Richards, Vyvyan, *Portrait of T.E. Lawrence* (Jonathan Cape, 1936)

Ricks, Christopher, *Tennyson* (Macmillan, 1989)

Rimbaud, Penny aka J. J. Ratter, *Shibboleth: My Revolting Life* (AK Press, 1998)

Rimbaud, Penny aka J. J. Ratter, *The Last of the Hippies: An Hysterical Romance* (AK Press, 2009)

Rose, June, *Daemons and Angels: A Life of Jacob Epstein* (Constable, 2002)

Scott, Jeff, *Showered in Shale* (Methanol, 2006)

Sharpe, James, *Dick Turpin: The Myth of the English Highwayman* (Profile, 2004)

Sharr, Adam, *Heidegger's Hut* (MIT Press, 2006)

Shaxson, Nicholas, *Treasure Islands: Tax Havens and the Men Who Stole the World* (Bodley Head, 2011)

Shenk, David, *The Forgetting* (Doubleday, 2001)

Sinclair, Iain, *Edge of the Orison* (Hamish Hamilton, 2005)

Sjöö, Monica and Mor, Barbara, *The Great Cosmic Mother* (Harper & Row, 1987)

Stone, C. J., *Fierce Dancing* (Faber & Faber, 1996)

Stone, C. J. and Pendragon, Arthur, *The Trials of Arthur: Revised Edition* (Big Hand, 2012)

Tate, W. E., *The English Village Community and the Enclosure Movements* (Gollancz, 1967)

Thompson, E. P., *The Making of the English Working Class* (Gollancz, 1963)

Thompson, E. P., *Whigs and Hunters: The Origin of the Black Act* (Allen Lane, 1975)

Tibble, J. W. and Tibble, Anne, *John Clare: A Life* (Joseph, 1972)

Tyme, John, *Motorways versus Democracy* (Macmillan, 1978)

Waller, Gary, *The Sidney Family Romance* (Wayne State University Press, 1993)

Williams, Merryn and Raymond (eds.), *John Clare: Selected Poetry and Prose* (Methuen, 1986)

Woodbine, George E., 'Origins of the Action of Trespass' (in the *Yale Law Journal*, June 1924)

Wroth, Mary, *The Countess of Montgomery's Urania* (1621), abridged and edited by Mary Ellen Lamb (Arizona Center for Medieval and Renaissance Studies, 2011)

Wroth, Mary, *The Poems of Lady Mary Wroth*, edited and introduced by Josephine A. Roberts (Louisiana State University Press, 1983)

Yates, Frances, *The Occult Philosophy in the Elizabethan Age* (Routledge & Kegan Paul, 1979)

Yates, Frances, *The Rosicrucian Enlightenment* (Routledge & Kegan Paul, 1972)

Acknowledgements

If writing is supposed to be a solitary pursuit, I'm obviously doing it wrong.

Thanks to the many people who read the various drafts of this book and encouraged or criticised, but always tried to help: Matthew Shapland and Nick Midgley (both of whom put up with me blah-ing on about it throughout the writing process), Simon Skevington, Jamie Collinson, Leila Baker, Adam Wishart, Alastair Siddons, Paul Carlin (with gratitude for the last-minute books too), Becky Thomas, Sean Preston, Will Atkins, John Ash, Carrie Plitt.

Thanks to my repositories of local knowledge: Simon Snell, Clare Shaw, Paul Holloway.

Thanks to the repositories of national knowledge: the staff of the British Library and, in particular, of the Rare Books and Music Reading Room.

Thanks to my agent, Patrick Walsh, my editor, Max Porter (young crow harrying aged magpie), Christine Lo, Sarah Wasley, Jenny Page, Pru Rowlandson and everyone at Granta.

Thanks to the people who offered me nudges, guidance, leg-ups and shortcuts even though they didn't have to: Geoff Garrison, Michael Lobban, Peter Linebaugh, Neil Goodwin, Ali Todd, Ben Thompson, Paul Ewen, Stanley Baker (for the

Pevsner), Louise Ashon, Laurence Bell, Nell Dunn and Lee Rourke.

Thanks, in particular, to everyone who so generously gave up their time, knowledge, opinions and memories in speaking to me: Penny Rimbaud, Gee Vaucher, Nikki Roberts, Susanna Lafond, Big John, Russell Houghton, Judy and Peter Adams, Ronna Gummer, Kerry Rolison.

Thanks to all the giants—and ordinary-size people—whose shoulders I've trodden on. I hope that if any small area of this book has piqued your interest you will be able to expand upon what's here using the notes and bibliography. However, I would particularly like to mention, once again, *Forests: The Shadow of Civilization* by Robert Pogue Harrison—a remarkable book which deserves to be much more widely read and which was key to the best of my thinking (while innocent of my many false turns).

Thanks, lastly, to my family—Leila, Miriam and Saul—for coming out to the woods with me, and for leaving me there, too.

Index

Keep in touch with
Granta Books:

Visit granta.com to discover more.

GRANTA